T0140952

Discrete Tomography of Delone Sets with Long-Range Order

Christian Huck

Bibliografische Information der Deutschen Nationalbibliothek

Die Deutsche Nationalbibliothek verzeichnet diese Publikation in der Deutschen Nationalbibliografie; detaillierte bibliografische Daten sind im Internet über http://dnb.d-nb.de abrufbar.

ISBN 978-3-8325-1656-7

Logos Verlag Berlin
Comeniushof, Gubener Str. 47,
10243 Berlin
Tel.: +49 030 42 85 10 90
Fax: +49 030 42 85 10 92
INTERNET: http://www.logos-verlag.de

To Svenja

Discrete Tomography of Delone Sets with Long-Range Order

Dissertation zur Erlangung des Doktorgrades
an der
Fakultät für Mathematik
der
Universität Bielefeld

vorgelegt von
Christian Huck

Fakultät für Mathematik
Universität Bielefeld

April 2007

1. Berichterstatter: Prof. Dr. Michael Baake
2. Berichterstatter: Dr. Uwe Grimm

Datum der mündlichen Prüfung: 16.07.2007

Gedruckt auf alterungsbeständigem Papier °° ISO 9706.

Acknowledgements
I wish to thank my advisor Michael Baake for suggesting the topic of discrete tomography and for sharing his remarkable intuition with me. I am grateful for his guidance and his support in preparing this thesis and, in particular, for the opportunity to meet the leading experts both in the mathematics of aperiodic order and in discrete tomography. Moreover, I would like to thank Peter Gritzmann, Barbara Langfeld and Katja Lord (Center of Mathematical Sciences, University of Technology, Munich, Germany) for their cooperation, various fruitful discussions and suggestions. I also would like to thank Richard J. Gardner (Department of Mathematics, Western Washington University, Bellingham, USA), Uwe Grimm (Department of Mathematics, Faculty of Mathematics and Computing, The Open University, Milton Keynes, UK) and Peter A. B. Pleasants (Department of Mathematics, University of Queensland, Brisbane, Australia) for valuable discussions, comments and, in particular, for carefully reading parts of the manuscript and for suggesting various simplifications. Further, I wish to thank the Collaborative Research Centre 701 'Spectral Structures and Topological Methods in Mathematics', with Friedrich Götze as its speaker, for financial and ideal support. It is a pleasure to thank my fellow students Guido Elsner and Bernd Sing for accompanying me during my studies and for discussing various mathematical problems with me. Further, I wish to thank the entire research group, in particular Dirk Frettlöh, Christoph Richard, Uwe Schwerdtfeger and Peter Zeiner, for the stimulating atmosphere. Last but not least, I would like to thank my family for their support, and my sister Katharina in particular.

Contents

Introduction

Tomography (derived from the Greek word $\tau o \mu o \sigma$ = slice) is concerned with the inverse problem of retrieving information about some object from (generally noisy) information about its slices. More concretely, in order to describe the problem of reconstructing a planar object (e.g., a slice of an object in Euclidean 3-space, given by a non-negative continuous function $F : \mathbb{R}^3 \longrightarrow \mathbb{R}$ with compact support) from its *1-dimensional continuous X-ray images*, consider a non-negative continuous function $f : \mathbb{R}^2 \longrightarrow \mathbb{R}$ with compact support. Further, for a *direction* $u \in \mathbb{S}^1$ (the unit circle), let the *1-dimensional X-ray transform of f* be the function $X_{(.)}f(.)\colon \{(u,z) \,|\, u \in \mathbb{S}^1,\, z \in (\mathbb{R}u)^\perp\} \longrightarrow \mathbb{R}$, defined by

$$X_u f(z) \;:=\; \int_{-\infty}^{+\infty} f(z+tu)\,dt\,.$$

Then, for $u \in \mathbb{S}^1$, the induced function $X_u f(.) : (\mathbb{R}u)^\perp \longrightarrow \mathbb{R}$ is also called the (*parallel*) *X-ray of f in direction u*. Moreover, the map R, defined by assigning to each non-negative continuous function $f : \mathbb{R}^2 \longrightarrow \mathbb{R}$ with compact support the map $X_{(.)}f(.)$, is also called the *Radon transform*. For the reconstruction problem, one has to reconstruct the function f from its X-rays in several directions u. Radon showed that f is uniquely determined by its X-rays in *all* directions $u \in \mathbb{S}^1$. Moreover, he proved an inversion formula in this case; see [**55**]. Although the result of Radon theoretically solves the reconstruction problem from above, it is of limited use in practice, since there one is only given X-ray information in a *finite* number of directions and, moreover, since there is always some noise involved when physical measurements are taken, one is confronted with the problem of imprecise data. Since Lorentz has shown in [**47**] that, given a finite set $U \subset \mathbb{S}^1$ of directions, there are always two different objects in the plane with the same X-rays in the directions of U, the above problem of uniquely reconstructing a planar object from its X-ray images actually is an *ill-posed* problem. Nevertheless, *Computerized Tomography* (CT) possesses efficient algorithms that handle such problems successfully. The most prominent among these algorithms are the *Filtered Back-Projection Algorithms*, which are based on the *Fourier Slice Theorem* (also known as the *Projection Slice Theorem*); see [**50**].

Discrete tomography[1] (DT) is, in contrast to the continuous setting above, concerned with the inverse problem of retrieving information about some *finite* object, e.g., given by a function $f\colon \mathbb{R}^3 \longrightarrow \{0,1\}$ with finite support, from (generally noisy) information about its slices. Here, a typical example is the *reconstruction* of a finite point set from its line sums in a small number of directions. More precisely, for the discrete analogue of 1-dimensional

[1]The name *discrete tomography* was introduced by Larry Shepp, who organized the first meeting on this topic at DIMACS on September 19, 1994.

continuous X-rays, a (*discrete parallel*) *X-ray* of a finite subset of Euclidean d-space $\mathbb{R}^d$ in direction u gives the number of points in the set on each line in $\mathbb{R}^d$ parallel to u. In the classical setting, motivated by crystals, the positions to be determined live on the square lattice $\mathbb{Z}^2$ or, more generally, on arbitrary lattices L in $\mathbb{R}^d$, where $d \geq 2$. Here, a subset S of $\mathbb{R}^d$ is said to *live on* a subgroup G of $\mathbb{R}^d$ when its difference set $S - S := \{s - s' \mid s, s' \in S\}$ is a subset of G. In fact, many of the problems in discrete tomography have been studied on $\mathbb{Z}^2$, the classical planar setting of discrete tomography; see [**41**], [**39**] and [**35**]. In the longer run, by also having other structures than perfect crystals in mind, one has to take into account wider classes of sets, or at least significant deviations from the lattice structure. As an intermediate step between periodic and random (or amorphous) Delone sets (defined below), we consider Delone sets with *long-range order*, thus including systems of *aperiodic order*, more precisely, of so-called *model sets* (or *mathematical quasicrystals*), which are commonly accepted as a good mathematical model for quasicrystalline structures in nature; see [**65**]. A particularly interesting question is to know whether these sets behave like lattices as far as discrete tomography is concerned, or whether they are closer to generic sets.

The main motivation for our interest in the discrete tomography of model sets comes from the physical existence of quasicrystals that can be described as aperiodic model sets together with the demand of materials science to reconstruct three-dimensional (quasi)crystals or planar layers of them from their images under quantitative *high resolution transmission electron microscopy* (HRTEM) in a small number of directions. In fact, in [**43, 61**] a technique is described, based on HRTEM, which can effectively measure the number of atoms lying on lines parallel to certain directions; it is called QUANTITEM (**QU**antitative **AN**alysis of **T**he **I**nformation from **T**ransmission **E**lectron **M**icroscopy). At present, the measurement of the number of atoms lying on a line can only be achieved for some crystals; see [**43, 61**] again. However, it is reasonable to expect that future developments in technology will improve this situation.

Of course, the continuous methods described above do not seem appropriate anymore. One reason for this is that, on the one hand, general CT reconstruction methods require X-ray data in a few hundred directions and, on the other hand, in DT only X-rays in a small number of directions are allowed since after about 3 to 5 images taken by HRTEM, typical objects may be damaged or even destroyed by the radiation energy. Instead, for discrete tomography, the methods needed come from combinatorial optimization, number theory, geometry, combinatorics and stochastics; see [**39**] for a summary of known results in the classical planar case mentioned above.

Whether or not one has future applications in materials science in mind, the starting point will always be a specific structure model. This means that the specific type of the (quasi)crystal is given (see [**5, 62**] for the concept), and one is confronted with the X-ray data of an unknown finite subset of it. This is the correct analogue of starting with translates of $\mathbb{Z}^d$ in the classical setting.

In this thesis, the investigation of the discrete tomography of systems with aperiodic order will be restricted to the study of a class of two-dimensional systems, the *cyclotomic model sets*, and one specific class of three-dimensional systems, the *icosahedral model sets.* In particular, we shall present one instance of the fact that model sets possess, as it is the case for lattices, a dimensional hierarchy, meaning that any model set in d dimensions can be sliced into model sets of dimension $d-1$. Namely, in Corollary 1.102 it will be shown that icosahedral model sets can be sliced into translates of certain cyclotomic model sets, the latter being in full analogy to the fact that the cubic lattice $\mathbb{Z}^3$ can be sliced into translates of the square lattice, $\mathbb{Z}^2$. Clearly, solving the problems of discrete tomography for two-dimensional systems with aperiodic order thus lies at the heart of solving the corresponding problems in three dimensions.

The three chapters of this thesis are organized as follows. In the first chapter, apart from several technical results dealing with properties of cyclotomic fields, we mainly introduce the objects and problems of discrete tomography of model sets, i.e., by using the Minkowski representation of algebraic number fields, we introduce, for $n \notin \{1,2\}$, the corresponding class of *cyclotomic model sets* $\varLambda \subset \mathbb{C} \cong \mathbb{R}^2$ which live on $\mathbb{Z}[\zeta_n]$, where ζ_n is a primitive nth root of unity in $\mathbb{C}$, e.g., $\zeta_n = e^{2\pi i/n}$. The $\mathbb{Z}$-module $\mathbb{Z}[\zeta_n]$ is the ring of integers in the nth cyclotomic field $\mathbb{Q}(\zeta_n)$, and, for $n \notin \{1,2,3,4,6\}$, when viewed as a subset of the plane, is dense, whereas cyclotomic model sets $\varLambda$ are *Delone sets*, i.e., they are uniformly discrete and relatively dense. In fact, model sets are even *Meyer sets*, meaning that also $\varLambda - \varLambda$ is uniformly discrete; see [**48**]. It turns out that, except the cyclotomic model sets living on $\mathbb{Z}[\zeta_n]$ with $n \in \{3,4,6\}$ (these are exactly the translations of the square and the triangular lattice, respectively), cyclotomic model sets $\varLambda$ are *aperiodic*, meaning that they have no translational symmetries at all. Well-known examples are the planar model sets with N-fold cyclic symmetry that are associated with the square tiling ($n = N = 4$), the triangle tiling ($2n = N = 6$), the Ammann-Beenker tiling ($n = N = 8$), the Tübingen triangle tiling ($2n = N = 10$) and the shield tiling ($n = N = 12$). Note that orders $5, 8, 10$ and 12 occur as standard cyclic symmetries of genuine quasicrystals [**65**], which are basically stacks of planar aperiodic layers. Further, as an example of three-dimensional systems of aperiodic order, we introduce the class of *icosahedral model sets* $\varLambda \subset \mathbb{R}^3$, which are living on the set of imaginary parts of elements of the icosian ring $\mathbb{I}$.

In the second chapter, we investigate the *uniqueness problem* of discrete tomography, i.e., the (unique) *determination* of finite subsets of a fixed Delone set $\varLambda \subset \mathbb{R}^d$, where $d \geq 2$, by X-rays in a small number of suitably prescribed directions. For practical reasons, only X-rays in $\varLambda$-directions, i.e., directions parallel to non-zero elements of the difference set $\varLambda - \varLambda$ of $\varLambda$, are permitted. This uniqueness problem is motivated by the fact that the above *reconstruction problem* of discrete tomography can possess rather different solutions. More precisely, we shall say that a subset $\mathcal{E}$ of the set of all finite subsets of Euclidean d-space, where $d \geq 2$, is determined by the X-rays in a finite set U of directions if different sets F and F' in $\mathcal{E}$ have different X-rays, i.e., if there is a direction $u \in U$ such that the X-rays of F and F' in direction u differ. Without the restriction to $\varLambda$-directions, one can prove that the finite subsets of a fixed Delone set $\varLambda \subset \mathbb{R}^d$, where $d \geq 2$, can be determined by one X-ray (Proposition 2.1). In fact, any X-ray in a non-$\varLambda$-direction is suitable for this purpose. However, in practice (HRTEM), X-rays in non-$\varLambda$-directions are meaningless since the resolution coming from such X-rays would not be good enough to allow a quantitative analysis – neighbouring lines are not sufficiently separated. Proposition 2.5 demonstrates that the finite sets F of cardinality less than or equal to some $k \in \mathbb{N}$ in a fixed Delone set set $\varLambda \subset \mathbb{R}^d$, where $d \geq 2$, are determined by any set of $k+1$ X-rays in pairwise non-parallel $\varLambda$-directions. As we pointed out above, in practice one is interested in the determination of finite sets by X-rays in a small number of directions. Observing that the typical atomic structures to be determined comprise about 10^6 to 10^9 atoms, one realizes that the last result is not practical at all. In fact, it is also shown that, at least for so-called *algebraic Delone sets* (e.g., cyclotomic model sets) and icosahedral model sets, any fixed number of X-rays in Λ-directions is insufficient to determine the entire class of finite subsets of the set Λ (Proposition 2.3 and Remark 2.4).

In view of this, one realizes that it is necessary to impose some restriction in order to obtain positive uniqueness results.

As a first option, we restrict the set of finite subsets of a fixed Delone set $\Lambda \subset \mathbb{R}^d$, where $d \geq 2$, under consideration, i.e., we consider for every $R > 0$ the class of *bounded* subsets of Λ with diameter less than R. Since Λ is uniformly discrete, bounded subsets of Λ are finite. It is shown that, for all $R > 0$ and for all Delone sets $\Lambda \subset \mathbb{R}^d$ of *finite local complexity* (the latter meaning that the difference set $\Lambda - \Lambda$ is closed and discrete), where $d \geq 2$, there are two non-parallel prescribed Λ-directions such that the set of bounded subsets of Λ with diameter less than R is determined by the X-rays in these directions (Theorem 2.8). As special cases, the last result applies to arbitrary model sets $\Lambda \subset \mathbb{R}^d$ with $d \geq 2$ (Corollary 2.10), thereby also including the cases of translates of lattices L in $\mathbb{R}^d$, $d \geq 2$. Moreover, even the corresponding results for orthogonal projections on orthogonal complements of 1-dimensional (respectively, $(d-1)$-dimensional) subspaces generated by elements of $\Lambda - \Lambda$ (resp., $L - L = L$) hold; the multiplicity information coming along with X-rays is not needed. Unfortunately, these results are limited in practice because, in general, one cannot make sure that all the directions which are used yield images of high enough resolution.

As a second option, we restrict the set of finite subsets of a fixed algebraic Delone set Λ by considering the class of *convex subsets of* Λ. They are finite sets $C \subset \Lambda$ whose convex hulls contain no new points of Λ, i.e., finite sets $C \subset \Lambda$ with $C = \mathrm{conv}(C) \cap \Lambda$. It is shown that, for the special case of cyclotomic model sets, there are four pairwise non-parallel prescribed Λ-directions such that the set of convex subsets of any cyclotomic model set Λ living on $\mathbb{Z}[\zeta_n]$ is determined by the X-rays in these directions (Theorem 2.54(a)). It is further shown that, also in the general situation of algebraic Delone sets, three pairwise non-parallel Λ-directions never suffice for this purpose (Theorem 2.54(b) and Theorem 2.31(b)). Furthermore, it is shown that, for cyclotomic model sets, *any* seven pairwise non-parallel Λ-directions which satisfy a certain relation (see below for details) have the property that the set of convex subsets of any cyclotomic model set Λ is determined by the X-rays in these directions (Theorem 2.54(c)). It is also demonstrated that, in a sense, the number seven cannot be reduced to six in the latter result (Theorem 2.54(c)). These results heavily depend on a massive restriction of the set of Λ-directions. We shall also remove this restriction and present explicit results in the case of the important subclass of cyclotomic model sets Λ with co-dimension 2, i.e., cyclotomic model sets living on $\mathbb{Z}[\zeta_n]$, where $n \in \{5, 8, 10, 12\}$. Here, by using p-adic valuations, we are able to present a more suitable analysis which shows that uniqueness will be provided by any set of four Λ-directions whose slopes (suitably ordered) yield a cross ratio that does not map under the field norm of the field extension $\mathbb{Q}(\zeta_n + \bar{\zeta}_n)/\mathbb{Q}$ to a certain *finite* set of rational numbers (Theorem 2.56 and Example 2.57). (It will turn out that these cross ratios are indeed elements of the maximal real subfield $\mathbb{Q}(\zeta_n + \bar{\zeta}_n)$ of the nth cyclotomic field $\mathbb{Q}(\zeta_n)$.) It is also shown that, by the same analysis, similar results can be obtained for arbitrary algebraic Delone sets (e.g., cyclotomic model sets) (Theorem 2.31 and Theorem 2.59). In the case of cyclotomic model sets with co-dimension 2, we shall be able to present sets of four Λ-directions which provide uniqueness and, additionally, yield images of rather high resolution in the quantitative HRTEM of the corresponding (aperiodic) cyclotomic model sets, the latter making these results look promising. It is also shown how the above results can be used to obtain corresponding results for the class of icosahedral model sets $\Lambda \subset \mathbb{R}^3$ (Theorem 2.67 and Theorem 2.69).

A major task in achieving the above results involves examining so-called *U-polygons* in algebraic Delone sets Λ, which exhibit a weak sort of regularity. In the context of U-polygons in cyclotomic model sets, the question for affinely regular polygons in the plane with all their vertices in a fixed cyclotomic model set Λ arises rather naturally. Chrestenson [**22**] has shown that any (planar) regular polygon whose vertices are contained in $\mathbb{Z}^d$ for some $d \geq 2$ must have $3, 4$ or 6 vertices. More generally, Gardner and Gritzmann [**33**] have characterized the *affinely* regular lattice polygons, i.e., images of non-degenerate regular polygons under a non-singular affine transformation $\Psi : \mathbb{R}^2 \longrightarrow \mathbb{R}^2$ whose vertices are contained in $\mathbb{Z}^2$. It turned out that these are precisely the affinely regular triangles, parallelograms and hexagons. Clearly, their result remains valid when one replaces the square lattice $\mathbb{Z}^2$ by the triangular lattice $\mathbb{Z}[\zeta_3]$, since the latter is the image of the first under an invertible linear transformation of the plane. Observing that both the square and triangular lattice are examples of rings of integers $\mathbb{Z}[\zeta_n]$ in cyclotomic fields $\mathbb{Q}(\zeta_n)$, one can ask for a generalization of the above result to all these objects, viewed as $\mathbb{Z}$-modules in the plane. Using standard results from algebra and algebraic number theory, we provide a characterization in terms of a simple divisibility condition (Theorem 2.33). Moreover, we show that our characterization of affinely regular polygons remains valid when restricted to the corresponding class of cyclotomic model sets Λ (Corollary 2.36).

As a third option, we consider the interactive technique of *successive determination*, introduced by Edelsbrunner and Skiena [**28**] for continuous X-rays, in the case of (aperiodic) cyclotomic model sets, in which the information from previous X-rays may be used in deciding the direction for the next X-ray. Here, by using Minkowski representations of orders of algebraic number fields, it is shown that the finite subsets of any cyclotomic model set Λ can be successively determined by X-rays in two non-parallel Λ-directions. In fact, it is shown that, for $n \notin \{1, 2\}$, the finite subsets of the ring of cyclotomic integers $\mathbb{Z}[\zeta_n]$ can be successively determined by X-rays in two non-parallel $\mathbb{Z}[\zeta_n]$-directions, i.e., directions parallel to a non-zero element of $\mathbb{Z}[\zeta_n]$ (Theorem 2.85). Again, the corresponding results do even hold for orthogonal projections on orthogonal complements of 1-dimensional subspaces generated by elements of $\Lambda - \Lambda$; the multiplicity information coming along with X-rays is again not needed. Again, it is also shown how the above results can be used to obtain corresponding results for the class of icosahedral model sets $\Lambda \subset \mathbb{R}^3$ (Corollary 2.94). Unfortunately, these results are again limited in practice because, in general, one cannot make sure that all the directions which are used yield images of high enough resolution.

Previous studies have focussed on the 'anchored' case that the underlying ground set is located in a linear space, i.e., in a space with a specified location of the origin. The X-ray data is then taken with respect to this localization. Note that the rotational orientation of a (quasi)crystalline probe in an electron microscope can rather easily be done in the diffraction mode, prior to taking images in the high-resolution mode, though, in general, a natural choice of a translational origin is *not* possible. This problem has its origin in the practice of quantitative HRTEM since, in general, the X-ray information does not allow us to locate the examined sets. Hence, we believe that, in order to obtain applicable results, one has to deal with this 'non-anchored' case. It is shown in this thesis that, at least in an approximative sense, this case is often feasible both for cyclotomic and icosahedral model sets (Theorem 2.61, Remark 2.64, Remark 2.68, Corollary 2.89 and Corollary 2.96).

In the third chapter, the algorithmic problems of discrete tomography of model sets are discussed in the case of cyclotomic and icosahedral model sets. In particular, it is shown that, in the real RAM-model of computation, the problem of reconstructing finite subsets of model sets Λ given X-rays in *two* Λ-directions can be solved in polynomial time for a large class of cyclotomic and icosahedral model sets Λ (Theorem 3.28 and Theorem 3.33).

CHAPTER 1

Basics and Preparations

The main objects and problems of discrete tomography of Delone sets to be studied in this thesis are introduced. In particular, the important classes of algebraic Delone sets, cyclotomic model sets, and icosahedral model sets are described. Also, the necessary technical results are provided, most of them being concerned with cyclotomic fields.

1.1. Algebraic background, definitions, and notation

1.1.1. Notation. Natural numbers are always assumed to be positive, i.e.,

$$\mathbb{N} = \{1, 2, 3, \dots\}.$$

Throughout the text, we use the convention that the symbol $\subset$ includes equality. We denote the norm in Euclidean d-space $\mathbb{R}^d$ by $\|\cdot\|$. The unit sphere in $\mathbb{R}^d$ is denoted by $\mathbb{S}^{d-1}$, i.e., $\mathbb{S}^{d-1} = \{x \in \mathbb{R}^d \mid \|x\| = 1\}$. Moreover, the elements of $\mathbb{S}^{d-1}$ are also called *directions*. For $x \in \mathbb{R}^d \setminus \{0\}$, we denote by u_x the direction $x/\|x\| \in \mathbb{S}^{d-1}$. If $x \in \mathbb{R}$, then $\lfloor x \rfloor$ denotes the greatest integer less than or equal to x. For $r > 0$ and $x \in \mathbb{R}^d$, $B_r(x)$ is the open ball of radius r about x. If $k, l \in \mathbb{N}$, then (k, l) and $[k, l]$ denote their greatest common divisor and least common multiple, respectively. There should be no confusion with the usual notation of open (resp., closed) and bounded real intervals. For a subset $S \subset \mathbb{R}^d$, $k \in \mathbb{N}$ and $R > 0$, we denote by $\mathrm{card}(S)$, $\mathcal{F}(S)$, $\mathcal{F}_{\leq k}(S)$, $\mathcal{D}_{<R}(S)$, $\mathrm{int}(S)$, $\mathrm{cl}(S)$, $\mathrm{bd}(S)$, $\langle S \rangle_{\mathbb{Z}}$, $\langle S \rangle_{\mathbb{Q}}$, $\langle S \rangle_{\mathbb{R}}$, $\mathrm{conv}(S)$, $\mathrm{diam}(S)$ and $\mathbb{1}_S$ the cardinality, the set of finite subsets, the set of finite subsets of S having cardinality less than or equal to k, the set of subsets of S with diameter less than R, interior, closure, boundary, $\mathbb{Z}$-linear hull, $\mathbb{Q}$-linear hull, $\mathbb{R}$-linear hull, convex hull, diameter and characteristic function of S, respectively. The *dimension* of S is the dimension of its affine hull $\mathrm{aff}(S)$, and is denoted by $\dim(S)$. Further, a linear subspace T of $\mathbb{R}^d$ is called an *S-subspace* if it is generated by elements of the difference set

$$S - S := \{ s - s' \mid s, s' \in S \}$$

of S. A direction $u \in \mathbb{S}^{d-1}$ is called an *S-direction* if it is parallel to a non-zero element of $S - S$. If T is a linear subspace of $\mathbb{R}^d$, we denote the orthogonal projection of an element x of $\mathbb{R}^d$ on T by $x|T$. Moreover, we denote the orthogonal projection of a subset S of $\mathbb{R}^d$ on T by $S|T$. The orthogonal complement of T is denoted by $T^{\perp}$. If we use this notation with respect to other than the *canonical* inner product on $\mathbb{R}^d$, we shall say so explicitly. The symmetric difference of two sets A and B is $A \triangle B := (A \setminus B) \cup (B \setminus A)$. A subset S of $\mathbb{R}^d$ is said to *live on* a subgroup G of $\mathbb{R}^d$ when its difference set $S - S$ is a subset of G. Obviously, this is equivalent to the existence of a suitable $t \in \mathbb{R}^d$ such that $S \subset t + G$. As usual, $R^{\times}$ denotes the group of units of a given ring R. Finally, the *centroid* (or *centre of mass*) of an element

$F \in \mathcal{F}(\mathbb{R}^d)$ is defined as

$$\frac{1}{\operatorname{card}(F)}\Big(\sum_{f \in F} f\Big).$$

1.1.2. U-polygons.

DEFINITION 1.1. Consider the Euclidean plane, $\mathbb{R}^2$.

(a) A *linear transformation* (resp., *affine transformation*) $\Psi: \mathbb{R}^2 \longrightarrow \mathbb{R}^2$ is given by $z \longmapsto Az$ (resp., $z \longmapsto Az + t$), where A is a real 2×2 matrix and $t \in \mathbb{R}^2$. In both cases, Ψ is called *singular* when $\det(A) = 0$; otherwise, it is non-singular.

(b) A *homothety* $h: \mathbb{R}^2 \longrightarrow \mathbb{R}^2$ is given by $z \longmapsto \lambda z + t$, where $\lambda \in \mathbb{R}$ is positive and $t \in \mathbb{R}^2$. A homothety is called a *dilatation* if $t = 0$. Further, we call a homothety *expansive* if $\lambda > 1$.

DEFINITION 1.2. Let $S \subset \mathbb{R}^2$ and let $T \subset \mathbb{R}^d$, where $d \in \mathbb{N}$.

(a) A *convex polygon* is the convex hull of a finite set of points in $\mathbb{R}^2$.

(b) A *polygon in S* is a convex polygon with all vertices in S.

(c) A finite subset C of T is called a *convex subset of* T if its convex hull contains no new points of T, i.e., if $C = \operatorname{conv}(C) \cap T$. Moreover, the set of all convex subsets of T is denoted by $\mathcal{C}(T)$.

(d) A *regular polygon* is always assumed to be planar, non-degenerate and convex. An *affinely regular polygon* is a non-singular affine image of a regular polygon. In particular, it must have at least 3 vertices.

(e) Let $U \subset \mathbb{S}^1$ be a finite set of directions. We call a non-degenerate convex polygon P a *U-polygon* if it has the property that whenever v is a vertex of P and $u \in U$, the line ℓ_u^v in the plane in direction u which passes through v also meets another vertex v' of P.

REMARK 1.3. Note that U-polygons have an even number of vertices. Moreover, an affinely regular polygon with an even number of vertices is a U-polygon if and only if each direction of U is parallel to one of its edges.

1.1.3. An index theorem, PV-numbers, field norms, and extensions of isomorphisms. Recall that a free Abelian group of rank r is, as a group, isomorphic with the integer lattice $\mathbb{Z}^r$ in Euclidean space $\mathbb{R}^r$.

PROPOSITION 1.4. *If G is a torsion-free Abelian group of rank r, and H is a subgroup which is also of rank r, then the subgroup index $[G : H]$ is finite and equals the absolute value of the determinant of the transition matrix A from any $\mathbb{Z}$-basis of G to any $\mathbb{Z}$-basis of H.*

PROOF. See [**18**, Chapter 2, Lemma 6.1.1]. □

DEFINITION 1.5. Let λ be a real algebraic integer.

(a) We call λ a *Pisot-Vijayaraghavan* number (*PV-number*) if $\lambda > 1$ while all its (algebraic) conjugates have moduli strictly less than 1.

(b) We call λ a *Pisot-Vijayaraghavan unit* (*PV-unit*) if λ is both a PV-number and a unit (i.e., if $1/\lambda$ is an algebraic integer as well).

REMARK 1.6. Let $\mathbb{K}/\mathbb{k}$ be an extension of algebraic number fields, say $\mathbb{K}/\mathbb{k}$ is of degree $d := [\mathbb{K} : \mathbb{k}] \in \mathbb{N}$. The corresponding norm $N_{\mathbb{K}/\mathbb{k}} \colon \mathbb{K} \longrightarrow \mathbb{k}$ is given by

$$N_{\mathbb{K}/\mathbb{k}}(\kappa) = \prod_{j=1}^{d} \sigma_j(\kappa),$$

where the σ_j are the d distinct embeddings of $\mathbb{K}/\mathbb{k}$ into $\mathbb{C}/\mathbb{k}$; compare [**18**, Algebraic Supplement, Sec. 2, Corollary 1]. In particular, for every $\kappa \in \mathbb{k}$, one has $N_{\mathbb{K}/\mathbb{k}}(\kappa) = \kappa^d$. Moreover, the norm is transitive in the following sense. If $\mathbb{L}$ is any intermediate field of $\mathbb{K}/\mathbb{k}$ above, then one has

$$N_{\mathbb{K}/\mathbb{k}} = N_{\mathbb{L}/\mathbb{k}} \circ N_{\mathbb{K}/\mathbb{L}}. \tag{1.1}$$

LEMMA 1.7. *Let $\sigma \colon \mathbb{K} \longrightarrow \mathbb{K}'$ be an isomorphism of fields, let $\mathbb{E}$ be an algebraic extension of $\mathbb{K}$, and let $\mathbb{L}$ be an algebraically closed extension of $\mathbb{K}'$. Then, there exists a field homomorphism $\sigma' : \mathbb{E} \longrightarrow \mathbb{L}$ which extends σ.*

PROOF. See [**46**, Ch. V.2, Theorem 2.8]. □

1.1.4. Cyclotomic fields. For all $n \in \mathbb{N}$, and ζ_n a fixed primitive nth root of unity in $\mathbb{C}$ (e.g., $\zeta_n = e^{2\pi i/n}$), let $\mathbb{Q}(\zeta_n)$ be the corresponding cyclotomic field. It is well known that $\mathbb{Q}(\zeta_n + \bar{\zeta}_n)$ is the maximal real subfield of $\mathbb{Q}(\zeta_n)$; see [**66**]. Throughout this paper, we shall use the notation

$$\mathbb{K}_n = \mathbb{Q}(\zeta_n),\ \mathbb{k}_n = \mathbb{Q}(\zeta_n + \bar{\zeta}_n),\ \mathcal{O}_n = \mathbb{Z}[\zeta_n],\ \mathcal{o}_n = \mathbb{Z}[\zeta_n + \bar{\zeta}_n],$$

where $\bar{\cdot}$ denotes complex conjugation. Further, ϕ will always denote Euler's phi-function (often also called Euler's totient function), i.e.,

$$\phi(n) = \operatorname{card}\left(\{k \in \mathbb{N} \mid 1 \le k \le n \text{ and } (k,n) = 1\}\right).$$

Occasionally, we identify $\mathbb{C}$ with $\mathbb{R}^2$. Moreover, we denote the set of rational primes by $\mathbb{P}$ and its subset of Sophie Germain prime numbers (i.e., primes p for which the number $2p+1$ is prime as well) by $\mathbb{P}_{SG}$.

REMARK 1.8. Sophie Germain prime numbers $p \in \mathbb{P}_{SG}$ are the primes p such that the equation $\phi(n) = 2p$ has solutions. It is not known whether there are infinitely many Sophie Germain primes. The first few are

$$\{2, 3, 5, 11, 23, 29, 41, 53, 83, 89, 113, 131, 173, \\ 179, 191, 233, 239, 251, 281, 293, 359, 419, \dots\},$$

see entry A005384 of [**63**] for further details.

LEMMA 1.9. *For $n \ge 3$, one has:*

(a) *$\mathcal{O}_n$ is an $\mathcal{o}_n$-module of rank 2. More precisely, $\{1, \zeta_n\}$ is an $\mathcal{o}_n$-basis of $\mathcal{O}_n$.*
(b) *$\mathbb{K}_n$ is a $\mathbb{k}_n$-vector space of dimension 2. More precisely, $\{1, \zeta_n\}$ is a $\mathbb{k}_n$-basis of $\mathbb{K}_n$.*

PROOF. First, we show (a). The linear independence of $\{1, \zeta_n\}$ over $\mathcal{o}_n$ is clear (since by our assumption $n \ge 3$, $\{1, \zeta_n\}$ is even linearly independent over $\mathbb{R}$). It suffices to prove that all non-negative integral powers ζ_n^m satisfy $\zeta_n^m = \alpha + \beta\zeta_n$ for suitable $\alpha, \beta \in \mathcal{o}_n$. Using induction, we are done if we show $\zeta_n^2 = \alpha + \beta\zeta_n$ for suitable $\alpha, \beta \in \mathcal{o}_n$. To this end, note $\bar{\zeta}_n = \zeta_n^{-1}$ and observe that $\zeta_n^2 = -1 + (\zeta_n + \zeta_n^{-1})\zeta_n$. Claim (b) follows from the same argument. □

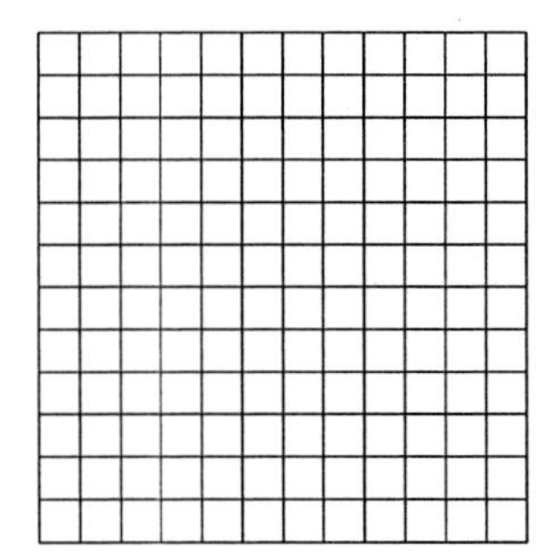

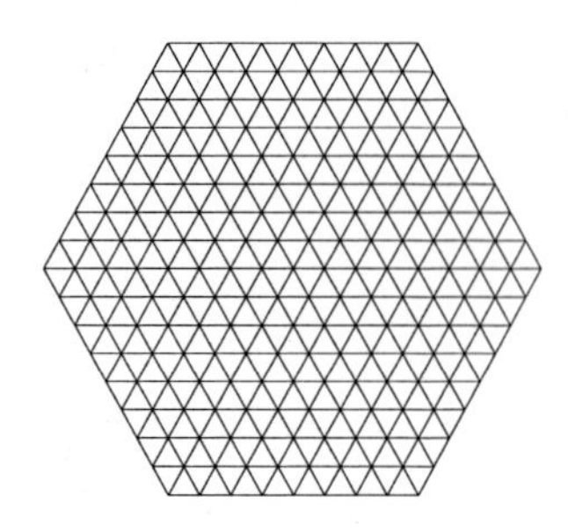

FIGURE 1.1. A central patch of the square tiling with vertex set $\varLambda_{\text{SQ}} = \mathcal{O}_4$ (left) and a central patch of the triangular tiling with vertex set $\varLambda_{\text{TRI}} = \mathcal{O}_3 = \mathcal{O}_6$ (right).

REMARK 1.10. Seen as a point set of $\mathbb{R}^2$, $\mathcal{O}_n$ has N-fold cyclic symmetry, where

$$N = N(n) := [n, 2]\,. \tag{1.2}$$

Except for the one-dimensional case ($n \in \{1, 2\}$, where $\mathcal{O}_1 = \mathcal{O}_2 = \mathbb{Z}$) and the crystallographic cases ($n \in \{3, 6\}$ (triangular lattice $\mathcal{O}_3 = \mathcal{O}_6$) and $n = 4$ (square lattice $\mathcal{O}_4$), see Figure 1.1), $\mathcal{O}_n$ is dense in $\mathbb{R}^2$. For the latter, note that, by Lemma 1.9, $\mathcal{O}_n$ is an $\mathcal{o}_n$-module of rank 2, whose $\mathbb{R}$-span is all of $\mathbb{R}^2$. For $n \in \mathbb{N} \setminus \{1, 2, 3, 4, 6\}$, $\mathcal{o}_n$ is a $\mathbb{Z}$-module of rank ≥ 2 (see Remark 1.24 below) embedded in $\mathbb{R}$, hence a dense set in $\mathbb{R}$. Consequently, $\mathcal{O}_n$ is then a dense set in $\mathbb{R}^2$.

PROPOSITION 1.11 (Gauß). *$[\mathbb{K}_n : \mathbb{Q}] = \phi(n)$, hence the set $\{1, \zeta_n, \zeta_n^2, \dots, \zeta_n^{\phi(n)-1}\}$ is a $\mathbb{Q}$-basis of $\mathbb{K}_n$. The field extension $\mathbb{K}_n/\mathbb{Q}$ is a Galois extension with Abelian Galois group $G(\mathbb{K}_n/\mathbb{Q}) \cong (\mathbb{Z}/n\mathbb{Z})^\times$, where $a \,(\text{mod}\, n)$ corresponds to the automorphism given by $\zeta_n \longmapsto \zeta_n^a$.*

PROOF. See [**66**, Theorem 2.5]. $\square$

REMARK 1.12. Note the identity $(\mathbb{Z}/n\mathbb{Z})^\times = \{a \,(\text{mod}\, n) \,|\, (a, n) = 1\}$ and see [**6**, Table 3] for examples of the explicit structure of $G(\mathbb{K}_n/\mathbb{Q})$. This reference also contains further material and references on the role of $\mathcal{O}_n$ in the context of model sets.

LEMMA 1.13 (Degree formula). *Let $E/F/K$ be an extension of fields. Then*

$$[E : K] = [E : F][F : K]\,.$$

PROOF. See [**46**, Ch. V.1, Proposition 1.2]. $\square$

COROLLARY 1.14. *If $n \geq 3$, one has $[\Bbbk_n : \mathbb{Q}] = \phi(n)/2$. Hence, a $\mathbb{Q}$-basis of $\Bbbk_n$ is given by the set $\{1, (\zeta_n + \bar{\zeta}_n), (\zeta_n + \bar{\zeta}_n)^2, \dots, (\zeta_n + \bar{\zeta}_n)^{\phi(n)/2-1}\}$. Moreover, $\Bbbk_n/\mathbb{Q}$ is a Galois extension with Abelian Galois group $G(\Bbbk_n/\mathbb{Q}) \cong (\mathbb{Z}/n\mathbb{Z})^\times/\{\pm 1 \,(\text{mod}\, n)\}$.*

PROOF. This is an immediate consequence of Lemma 1.9(b), Lemma 1.13, Proposition 1.11 and Galois theory. □

REMARK 1.15. The dimension statement of Corollary 1.14 also follows via Galois theory from Lemma 1.13 in connection with the fact that $\mathbb{k}_n$ is the fixed field of $\mathbb{K}_n$ with respect to the subgroup $\{id, \bar{\cdot}\}$ of $G(\mathbb{K}_n/\mathbb{Q})$, where $\bar{\cdot}$ denotes complex conjugation, i.e., the automorphism given by $\zeta_n \longmapsto \zeta_n^{-1}$ (recall that $\mathbb{k}_n$ is the maximal real subfield of $\mathbb{K}_n$).

LEMMA 1.16. *If* $m, n \in \mathbb{N}$, *then* $\mathbb{K}_m \cap \mathbb{K}_n = \mathbb{K}_{(m,n)}$.

PROOF. The assertion follows from similar arguments as in the proof of the special case $(m, n) = 1$; compare [**46**, Ch. VI.3, Corollary 3.2]. Here, one has to observe $\mathbb{Q}(\zeta_m, \zeta_n) = \mathbb{K}_m\mathbb{K}_n = \mathbb{K}_{[m,n]}$ and then to employ the identity

$$\phi(m)\phi(n) = \phi([m,n])\phi((m,n)) \tag{1.3}$$

instead of merely using the multiplicativity of the arithmetic function ϕ. □

LEMMA 1.17. *Let* $m, n \in \mathbb{N}$. *The following statements are equivalent:*

(i) $\mathbb{K}_m \subset \mathbb{K}_n$.
(ii) $m|n$, *or* $m \equiv 2 \pmod 4$ *and* $m|2n$.

PROOF. For direction (ii) $\Rightarrow$ (i), the assertion is clear if $m|n$. Further, if $m \equiv 2 \pmod 4$, say $m = 2o$ for a suitable odd number o, and $m|2n$, then $\mathbb{K}_o \subset \mathbb{K}_n$ (due to $o|n$). However, Proposition 1.11 shows that the inclusion of fields $\mathbb{K}_o \subset \mathbb{K}_{2o} = \mathbb{K}_m$ cannot be proper since we have, by means of the multiplicativity of ϕ, $\phi(m) = \phi(2o) = \phi(o)$. This gives $\mathbb{K}_m \subset \mathbb{K}_n$.

For direction (i) $\Rightarrow$ (ii), suppose $\mathbb{K}_m \subset \mathbb{K}_n$. Then, Lemma 1.16 implies $\mathbb{K}_m = \mathbb{K}_{(m,n)}$, hence

$$\phi(m) = \phi((m,n))\,; \tag{1.4}$$

see Proposition 1.11 again. Using the multiplicativity of ϕ together with $\phi(p^j) = p^{j-1}(p-1)$ for $p \in \mathbb{P}$ and $j \in \mathbb{N}$, we see that, given the case $(m, n) < m$, (1.4) can only be fulfilled if $m \equiv 2 \pmod 4$ and $m|2n$. The remaining case $(m, n) = m$ is equivalent to the relation $m|n$. □

COROLLARY 1.18. *Let* $m, n \in \mathbb{N}$. *The following statements are equivalent:*

(i) $\mathbb{K}_m = \mathbb{K}_n$.
(ii) $m = n$, *or* m *is odd and* $n = 2m$, *or* n *is odd and* $m = 2n$. □

REMARK 1.19. Corollary 1.18 implies that, for $m, n \not\equiv 2 \pmod 4$, one has the identity $\mathbb{K}_m = \mathbb{K}_n$ if and only if $m = n$.

LEMMA 1.20. *Let* $m, n \in \mathbb{N}$ *with* $m, n \geq 3$. *Then, one has:*

(a) $\mathbb{k}_m = \mathbb{k}_n \iff \mathbb{K}_m = \mathbb{K}_n$ *or* $m, n \in \{3, 4, 6\}$.
(b) $\mathbb{k}_m \subset \mathbb{k}_n \iff \mathbb{K}_m \subset \mathbb{K}_n$ *or* $m \in \{3, 4, 6\}$.

PROOF. For claim (a), let us suppose $\mathbb{k}_m = \mathbb{k}_n =: \mathbb{k}$ first. Then, Proposition 1.11 and Corollary 1.14 imply that $[\mathbb{K}_m : \mathbb{k}] = [\mathbb{K}_n : \mathbb{k}] = 2$. Note that $\mathbb{K}_m \cap \mathbb{K}_n = \mathbb{K}_{(m,n)}$ is a cyclotomic field containing $\mathbb{k}$. It follows from Lemma 1.13 that either $\mathbb{K}_m \cap \mathbb{K}_n = \mathbb{K}_{(m,n)} = \mathbb{K}_m = \mathbb{K}_n$ or $\mathbb{K}_m \cap \mathbb{K}_n = \mathbb{K}_{(m,n)} = \mathbb{k}$ and hence $\mathbb{k}_m = \mathbb{k}_n = \mathbb{k} = \mathbb{Q}$, since the latter is the only real cyclotomic field. Now, this implies $m, n \in \{3, 4, 6\}$; see also Lemma 1.21(a) below. The other direction is obvious. Claim (b) follows immediately from the part (a). □

LEMMA 1.21. *Consider ϕ on $\{n \in \mathbb{N} \mid n \not\equiv 2 \,(\mathrm{mod}\, 4)\}$. Then, one has:*

(a) *$\phi(n)/2 = 1$ if and only if $n \in \{3, 4\}$. These are the crystallographic cases of the plane.*
(b) *$\phi(n)/2 \in \mathbb{P}$ if and only if $n \in \mathbb{S} := \{8, 9, 12\} \cup \{2p + 1 \mid p \in \mathbb{P}_{SG}\}$.*

PROOF. The equivalences follow from the multiplicativity of ϕ in conjunction with the identity $\phi(p^j) = p^{j-1}(p-1)$ for $p \in \mathbb{P}$ and $j \in \mathbb{N}$. □

REMARK 1.22. Let $n \not\equiv 2 \,(\mathrm{mod}\, 4)$. By Corollary 1.14, $\mathbb{k}_n/\mathbb{Q}$ is a Galois extension with Abelian Galois group $G(\mathbb{k}_n/\mathbb{Q})$ of order $\phi(n)/2$. Now, it follows from Lemma 1.21 that $G(\mathbb{k}_n/\mathbb{Q})$ is trivial if and only if $n \in \{1, 3, 4\}$, and simple if and only if $n \in \mathbb{S}$, with $\mathbb{S}$ as defined in Lemma 1.21(b).

A full $\mathbb{Z}$-module (i.e., a module of full rank) in an algebraic number field $\mathbb{K}$ which contains the number 1 and is a ring is called an *order* of $\mathbb{K}$. It turns out that among the various orders of $\mathbb{K}$ there is one *maximal order* which contains all the other orders, namely the ring of integers in $\mathbb{K}$; see [**18**, Chapter 2, Section 2]. For cyclotomic fields, one has the following well-known result.

PROPOSITION 1.23. *For $n \in \mathbb{N}$, one has:*

(a) *$\mathcal{O}_n$ is the ring of cyclotomic integers in $\mathbb{K}_n$, and hence its maximal order.*
(b) *$\mathcal{o}_n$ is the ring of integers of $\mathbb{k}_n$, and hence its maximal order.*

PROOF. See [**66**, Theorem 2.6 and Proposition 2.16]. □

REMARK 1.24. It follows from Proposition 1.23(a) and Proposition 1.11 that $\mathcal{O}_n$ is a $\mathbb{Z}$-module of rank $\phi(n)$ with $\mathbb{Z}$-basis $\{1, \zeta_n, \zeta_n^2, \dots, \zeta_n^{\phi(n)-1}\}$. Likewise, Proposition 1.23(b) and Corollary 1.14 imply that $\mathcal{o}_n$ is a $\mathbb{Z}$-module of rank $\phi(n)/2$. Moreover, a $\mathbb{Z}$-basis of $\mathcal{o}_n$ is given by the set $\{1, (\zeta_n + \bar{\zeta}_n), (\zeta_n + \bar{\zeta}_n)^2, \dots, (\zeta_n + \bar{\zeta}_n)^{\phi(n)/2-1}\}$.

DEFINITION 1.25. The *nth cyclotomic polynomial* is given by

$$F_n := \prod_{\zeta} (X - \zeta),$$

where ζ runs over all primitive nth roots of unity in $\mathbb{C}$.

LEMMA 1.26. *For $n \in \mathbb{N}$, one has:*

(a) *F_n is monic and* $\deg(F_n) = \phi(n)$.
(b) $\prod_{d|n} F_d = X^n - 1$.
(c) $F_n \in \mathbb{Z}[X]$.

PROOF. See [**46**, Chapter VI.3]. □

REMARK 1.27. Lemma 1.26 shows that we can compute the nth cyclotomic polynomial recursively by use of the Euclidean algorithm in $\mathbb{Z}[X]$.

PROPOSITION 1.28 (Gauß). *The minimum polynomial* $\mathrm{Mipo}_{\mathbb{Q}}(\zeta_n)$ *of ζ_n over $\mathbb{Q}$ is the nth cyclotomic polynomial F_n.*

PROOF. By Definition 1.25, ζ_n is a root of F_n. Now, note that $\mathrm{Mipo}_{\mathbb{Q}}(\zeta_n)$ is, by definition, the (uniquely determined) monic polynomial in $\mathbb{Q}[X]$ of minimal degree having ζ_n as a root. Of course, it is a standard fact that $\deg(\mathrm{Mipo}_{\mathbb{Q}}(\zeta_n)) = [\mathbb{K}_n : \mathbb{Q}]$; see [**46**, Chapter V.1, Proposition 1.4]. By Proposition 1.11, one has $[\mathbb{K}_n : \mathbb{Q}] = \phi(n)$, hence the result follows from Lemma 1.26. □

LEMMA 1.29. *For all $n \in \mathbb{N}$ with $n \geq 3$, there is a PV-number of (full) degree $\phi(n)/2$ in $\mathcal{O}_n$.*

PROOF. This follows immediately from [**58**, Ch. I, Theorem 2]. □

REMARK 1.30. For $n \in \mathbb{N} \setminus \{1, 2, 3, 4, 6\}$, there is even a PV-unit in $\mathcal{O}_n$. This can be seen by considering a representation of the maximal order $\mathcal{O}_n$ of $\mathbb{k}_n$ (cf. Proposition 1.23(b)) in logarithmic space together with the fact that the units of $\mathcal{O}_n$ form a (full) lattice in the hyperplane given by

$$x_1 + \cdots + x_{\frac{\phi(n)}{2}} = 0;$$

see [**18**, Ch. 2, Sections 3 and 4]. Clearly, there are points of this lattice with $x_1 > 0$ and $x_2, \dots, x_{\phi(n)/2} < 0$, and the corresponding $\alpha \in \mathcal{O}_n$ with respect to this representation are PV-units. Note that these PV-units in $\mathcal{O}_n$ necessarily have (full) degree $\phi(n)/2$: if one of them had degree $(\phi(n)/2)/m$, its conjugates (under the effect of the Galois group $G(\mathbb{k}_n/\mathbb{Q})$, cf. Corollary 1.14) would come in sets of m equal conjugates and we could not have just one positive x (this is essentially Dirichlet's unit theorem). We also refer the reader to [**29**, Lemma 8.1.5(b)].

From now on, for $n \in \mathbb{N}$, we always let $\zeta_n := e^{2\pi i/n}$, as a specific choice for a primitive nth root of unity in $\mathbb{C}$.

1.1.5. A cyclotomic theorem. In this section, we need the following facts from the theory of p-adic valuations; compare [**38, 44**].

Let $p \in \mathbb{P}$. The p-adic valuation on $\mathbb{Z}$ is the function v_p, defined by $v_p(0) := \infty$ together with the equation

$$n = p^{v_p(n)} n'$$

for $n \neq 0$, where p does not divide n'; that is, $v_p(n)$ is the exponent of the highest power of p that divides n. The function v_p is extended to $\mathbb{Q}$ by defining

$$v_p\Big(\frac{a}{b}\Big) := v_p(a) - v_p(b)$$

for $a, b \in \mathbb{Z} \setminus \{0\}$; see [**38**, p. 23]. Note that v_p is $\mathbb{Z}$-valued on $\mathbb{Q} \setminus \{0\}$. As in [**38**, Ch. 5], v_p can further be extended to the algebraic closure $\mathbb{Q}_p^{\mathrm{alg}}$ of a field $\mathbb{Q}_p$, whose elements are called *p-adic numbers*, containing $\mathbb{Q}$. Note that $\mathbb{Q}_p^{\mathrm{alg}}$ contains the algebraic closure $\mathbb{Q}^{\mathrm{alg}}$ of $\mathbb{Q}$ and hence all algebraic numbers. On $\mathbb{Q}_p^{\mathrm{alg}} \setminus \{0\}$, v_p takes values in $\mathbb{Q}$, and satisfies

$$v_p(-x) = v_p(x), \tag{1.5}$$

$$v_p(xy) = v_p(x) + v_p(y), \tag{1.6}$$

$$v_p\Big(\frac{x}{y}\Big) = v_p(x) - v_p(y) \tag{1.7}$$

and

$$v_p(x+y) \;\geq\; \min\{v_p(x), v_p(y)\}\,, \tag{1.8}$$

compare also [**38**, p. 143].

PROPOSITION 1.31. *Let $p \in \mathbb{P}$ and let $r, s, t \in \mathbb{N}$. If r is not a power of p and $(r,s) = 1$, one has*

$$v_p(1-\zeta_r^s) = 0\,. \tag{1.9}$$

Otherwise, if $(p,s) = 1$, then

$$v_p(1-\zeta_{p^t}^s) = \frac{1}{p^{t-1}(p-1)}\,. \tag{1.10}$$

PROOF. See [**33**, Proposition 3.6]. □

DEFINITION 1.32. Let $k, m \in \mathbb{N}$ and let $p \in \mathbb{P}$. An mth root of unity ζ_m^k is called a *p-power root of unity* if there is a $t \in \mathbb{N}$ such that $\frac{k}{m} = \frac{s}{p^t}$ for some $s \in \mathbb{N}$ with $(p,s) = 1$.

REMARK 1.33. Note that an mth root of unity ζ_m^k is a p-power root of unity if and only if it is a primitive p^tth root of unity for some $t \in \mathbb{N}$.

LEMMA 1.34. *Let $k, t \in \mathbb{N}$ and $p \in \mathbb{P}$. Further, let $j, m \in \mathbb{N}$ with $(j,m) = 1$. Then, one has:*

(a) *ζ_m^k is a primitive p^tth root of unity if and only if $(\zeta_m^j)^k$ is a primitive p^tth root of unity.*

(b) *ζ_m^k is a p-power root of unity if and only if $(\zeta_m^j)^k$ is a p-power root of unity.*

PROOF. We first prove claim (a). Assume that $\frac{k}{m} = \frac{s}{p^t}$ for a suitable $s \in \mathbb{N}$ with $(p,s) = 1$. In particular, it follows that $p|m$ and, since $(j,m) = 1$, one has $(p,j) = 1$. Hence, $\frac{jk}{m} = \frac{js}{p^t}$ and $(p, js) = 1$. Conversely, assume $\frac{jk}{m} = \frac{s}{p^t}$ for a suitable $s \in \mathbb{N}$ with $(p,s) = 1$. Since $(j,m) = 1$, it follows that $j|s$, say $jl = s$ for a suitable $l \in \mathbb{N}$. Hence, $\frac{k}{m} = \frac{l}{p^t}$ and, moreover, $(p,l) = 1$. Claim (b) follows immediately from part (a). □

LEMMA 1.35. *Let $m, k \in \mathbb{N}$ and let $p \in \mathbb{P}$. If $\sigma \in G(\mathbb{K}_m/\mathbb{Q})$, then*

$$v_p(1-\zeta_m^k) = v_p\big(\sigma(1-\zeta_m^k)\big)\,.$$

PROOF. By Proposition 1.11, σ is given by $\zeta_m \longmapsto \zeta_m^j$, where $j \in \mathbb{N}$ satisfies $(j,m) = 1$. The assertion follows immediately from Proposition 1.31 in conjunction with Lemma 1.34. □

REMARK 1.36. Note that Lemma 1.35 is only one instance of the following more general fact. It is well known that, if $\mathbb{K}$ is a normal algebraic number field (i.e., a finite Galois extension of $\mathbb{Q}$) and if $\kappa \in \mathbb{K}$, then one has $v_p(\kappa) = v_p(\sigma(\kappa))$ for every $p \in \mathbb{P}$ and every $\sigma \in G(\mathbb{K}/\mathbb{Q})$; see [**18**, Ch. 3, Sec. 4, Prob. 7] and the parenthetical clause of the sentence following [**66**, Proposition 2.14] together with [**66**, Theorem 2.13] and [**66**, Proposition 2.14] itself.

DEFINITION 1.37. Let $m \geq 4$ be a natural number. We define

$$D'_m := \left\{ (k_1, k_2, k_3, k_4) \in \mathbb{N}^4 \;\middle|\; k_1, k_2, k_3, k_4 \leq m-1 \text{ and } k_1 + k_2 = k_3 + k_4 \right\},$$

together with its subset

$$D_m := \left\{ (k_1, k_2, k_3, k_4) \in \mathbb{N}^4 \;\middle|\; k_3 < k_1 \leq k_2 < k_4 \leq m-1 \text{ and } k_1 + k_2 = k_3 + k_4 \right\},$$

and define the function $f_m : D'_m \longrightarrow \mathbb{C}$ by

$$f_m\big((k_1, k_2, k_3, k_4)\big) := \frac{(1-\zeta_m^{k_1})(1-\zeta_m^{k_2})}{(1-\zeta_m^{k_3})(1-\zeta_m^{k_4})}. \tag{1.11}$$

LEMMA 1.38. *Let $m \geq 4$. The function f_m takes only values on the positive real axis, i.e., one has $f_m[D'_m] \subset \{x \in \mathbb{R} \mid x > 0\}$. Moreover, one has $f_m(d) > 1$ for all $d \in D_m$.*

PROOF. See the proof of [**33**, Lemma 3.1]. □

COROLLARY 1.39. *Let $m \geq 4$ and let $d \in D'_m$. Then, one has $f_m(d) \in \mathbb{k}_m$.*

PROOF. By Lemma 1.38, $f_m(d)$ is real. The assertion follows immediately from the fact that $f_m(d) \in \mathbb{K}_m$ and the fact that $\mathbb{k}_m$ is the maximal real subfield of $\mathbb{K}_m$. □

For our application to discrete tomography, we need to investigate the set

$$\Big(\bigcup_{m \geq 4} f_m[D_m] \Big) \cap \mathbb{k} \tag{1.12}$$

for arbitrary real algebraic number fields $\mathbb{k}$. Gardner and Gritzmann showed the following result, which deals with the smallest among all algebraic number fields, i.e., with $\mathbb{k} = \mathbb{Q}$.

THEOREM 1.40.

$$\Big(\bigcup_{m \geq 4} f_m[D_m] \Big) \cap \mathbb{Q} = \left\{ \tfrac{4}{3}, \tfrac{3}{2}, 2, 3, 4 \right\} =: N_1 .$$

Moreover, all solutions of $f_m(d) = q \in \mathbb{Q}$, where $m \geq 4$ and $d := (k_1, k_2, k_3, k_4) \in D_m$, are either given, up to multiplication of m and d by the same factor, by $m = 12$ and one of the following

(i) $d = (6, 6, 4, 8), q = \frac{4}{3}$; (ii) $d = (6, 6, 2, 10), q = 4$;
(iii) $d = (4, 8, 3, 9), q = \frac{3}{2}$; (iv) $d = (4, 8, 2, 10), q = 3$;
(v) $d = (4, 4, 2, 6), q = \frac{3}{2}$; (vi) $d = (8, 8, 6, 10), q = \frac{3}{2}$;
(vii) $d = (4, 4, 1, 7), q = 3$; (viii) $d = (8, 8, 5, 11), q = 3$;
(ix) $d = (3, 9, 2, 10), q = 2$; (x) $d = (3, 3, 1, 5), q = 2$;
(xi) $d = (9, 9, 7, 11), q = 2$;

or by one of the following

(xii) $d = (2k, s, k, k+s), q = 2,$ *where* $s \geq 2, m = 2s$ *and* $1 \leq k \leq \frac{s}{2}$;
(xiii) $d = (s, 2k, k, k+s), q = 2,$ *where* $s \geq 2, m = 2s$ *and* $\frac{s}{2} \leq k < s$.

PROOF. See [**33**, Lemma 3.8, Lemma 3.9 and Theorem 3.10]. □

In this section, we do not aim at full analogues of Theorem 1.40 in the case of the set (1.12) for real algebraic number fields $\Bbbk \neq \mathbb{Q}$. (The study of this interesting problem is work in progress.) Instead, we show that the elements of the set (1.12) satisfy certain conditions.

In fact, although we are only interested in the set (1.12), we shall investigate its superset

$$\Big(\bigcup_{m\geq 4} f_m[D'_m]\Big) \cap \Bbbk \tag{1.13}$$

for arbitrary real algebraic number fields $\Bbbk$. Every (real) algebraic number field $\Bbbk$ different from $\mathbb{Q}$ has some finite degree $e := [\Bbbk : \mathbb{Q}] \geq 2$. In particular, we are interested in the case $e = 2$. In the following, we denote certain degrees by e (resp., f). We wish to emphasize that this should not be confused with their common usage in algebraic number theory (where e denotes the ramification index and f denotes the residue class degree).

LEMMA 1.41. *Let $m \geq 4$, let $p \in \mathbb{P}$, and let $d \in D'_m$. If $\sigma \in G(\mathbb{K}_m/\mathbb{Q})$, then*

$$v_p\big(f_m(d)\big) = v_p\Big(\sigma\big(f_m(d)\big)\Big).$$

PROOF. The assertion follows from Lemma 1.35 and Equations (1.6) and (1.7). □

LEMMA 1.42. *Let $m \geq 4$ and let $d \in D'_m$. Then, for any prime factor $p \in \mathbb{P}$ of the numerator of the field norm $N_{\mathbb{Q}(f_m(d))/\mathbb{Q}}(f_m(d))$, one has*

$$v_p\Big(N_{\mathbb{Q}(f_m(d))/\mathbb{Q}}\big(f_m(d)\big)\Big) = e\, v_p\big(f_m(d)\big) \in \mathbb{N},$$

where $e := [\mathbb{Q}(f_m(d)) : \mathbb{Q}] \in \mathbb{N}$ is the degree of $f_m(d)$ over $\mathbb{Q}$.

PROOF. By Corollary 1.39, one has $f_m(d) \in \Bbbk_m \subset \mathbb{K}_m$. This implies the inclusion of fields

$$\mathbb{Q}\big(f_m(d)\big) \subset \Bbbk_m \subset \mathbb{K}_m\,. \tag{1.14}$$

The norm $N_{\mathbb{Q}(f_m(d))/\mathbb{Q}}\colon \mathbb{Q}\big(f_m(d)\big) \longrightarrow \mathbb{Q}$ of the field extension $\mathbb{Q}\big(f_m(d)\big)/\mathbb{Q}$ is given by

$$N_{\mathbb{Q}(f_m(d))/\mathbb{Q}}(q) = \prod_{j=1}^{e} \sigma_j(q)$$

for $q \in \mathbb{Q}\big(f_m(d)\big)$, where $\{\sigma_j \,|\, j \in \{1,\dots,e\}\}$ is the Galois group $G\big(\mathbb{Q}\big(f_m(d)\big)/\mathbb{Q}\big)$; see Remark 1.6 and note that the field extension $\mathbb{Q}\big(f_m(d)\big)/\mathbb{Q}$ is indeed a Galois extension. The latter follows immediately from Galois theory, since, by Proposition 1.11, the Galois extension $\mathbb{K}_m/\mathbb{Q}$ has an Abelian Galois group; cf. Proposition 1.11. By Relation (1.14), Lemma 1.7, and since $\mathbb{K}_m/\mathbb{Q}$ is a Galois extension, each field automorphism

$$\sigma_j \in G\big(\mathbb{Q}\big(f_m(d)\big)/\mathbb{Q}\big),$$

$j \in \{1,\dots,e\}$, can be extended to a field automorphism $\sigma'_j \in G(\mathbb{K}_m/\mathbb{Q})$. It follows that

$$N_{\mathbb{Q}(f_m(d))/\mathbb{Q}}\big(f_m(d)\big) = \prod_{j=1}^{e} \sigma'_j\big(f_m(d)\big).$$

Using the p-adic valuation v_p in conjunction with Equation (1.6) and Lemma 1.41, one gets

$$v_p\Big(N_{\mathbb{Q}(f_m(d))/\mathbb{Q}}\big(f_m(d)\big)\Big) = e\, v_p\big(f_m(d)\big) \in \mathbb{N},$$

which completes the proof. □

LEMMA 1.43. *Let $m \geq 4$ and let $d \in D'_m$. Then, for any prime factor $p \in \mathbb{P}$ of the numerator of $N_{\mathbb{Q}(f_m(d))/\mathbb{Q}}(f_m(d))$, the expression $\zeta_m^{k_j}$ in (1.11) is a p-power root of unity for at least one value of $j \in \{1,2\}$.*

PROOF. Let $e := [\mathbb{Q}(f_m(d)) : \mathbb{Q}] \in \mathbb{N}$ be the degree of $f_m(d)$ over $\mathbb{Q}$. By assumption, the numerator of $N_{\mathbb{Q}(f_m(d))/\mathbb{Q}}(f_m(d))$ has a prime factor, say $p \in \mathbb{P}$. Using Lemma 1.42, one gets

$$v_p\big(f_m(d)\big) = v_p\left(\frac{(1-\zeta_m^{k_1})(1-\zeta_m^{k_2})}{(1-\zeta_m^{k_3})(1-\zeta_m^{k_4})}\right) \in \frac{1}{e}\mathbb{N}.$$

The assertion now follows from Equation (1.6), Equation (1.7) and Proposition 1.31. □

The following result is a rather simple one. For convenience, we include a proof.

LEMMA 1.44. *Let $m \geq 4$ and let $d := (k_1, k_2, k_3, k_4) \in D'_m$. Then one has the equality of fields $\mathbb{Q}(f_m(d)) = \mathbb{Q}(f_m(d'))$ and, moreover,*

$$N_{\mathbb{Q}(f_m(d'))/\mathbb{Q}}\big(f_m(d')\big) = \Big(N_{\mathbb{Q}(f_m(d))/\mathbb{Q}}\big(f_m(d)\big)\Big)^{-1},$$

where $d' := (k_3, k_4, k_1, k_2) \in D'_m$. In particular, $f_m(d)$ and $f_m(d')$ have the same degree over $\mathbb{Q}$.

PROOF. Clearly, one has $d' := (k_3, k_4, k_1, k_2) \in D'_m$. Further, observing the identity $f_m(d') = 1/f_m(d)$ (by Lemma 1.38, one has $f_m(d) \neq 0$), one sees the equality of the fields $\mathbb{Q}(f_m(d)) = \mathbb{Q}(f_m(d'))$. The identity

$$\begin{aligned} N_{\mathbb{Q}(f_m(d'))/\mathbb{Q}}\big(f_m(d')\big) &= N_{\mathbb{Q}(f_m(d))/\mathbb{Q}}\big(f_m(d')\big) \\ &= N_{\mathbb{Q}(f_m(d))/\mathbb{Q}}\Big(\big(f_m(d)\big)^{-1}\Big) \\ &= \Big(N_{\mathbb{Q}(f_m(d))/\mathbb{Q}}\big(f_m(d)\big)\Big)^{-1} \end{aligned}$$

completes the proof. □

The following definition will be useful.

DEFINITION 1.45. Let $a \in \mathbb{N}$, let $S \subset \mathbb{R} \setminus \{0\}$, and let $\alpha \in \mathbb{Z}$. We set:

(a) $a^{\downarrow} := \{\frac{a}{b} \in \mathbb{Q} \,|\, b \in \mathbb{N} \text{ with } b < a \text{ and } (a,b) = 1\}$.

(b) $[S]^{\alpha} := \{x^{\alpha} \,|\, x \in S\}$.

1.1.5.1. *The case of degree two.*

LEMMA 1.46. *Let $m \geq 4$ and let $d := (k_1, k_2, k_3, k_4) \in D'_m$. Suppose that $f_m(d)$ is of degree two over $\mathbb{Q}$ and suppose that the absolute value of $N_{\mathbb{Q}(f_m(d))/\mathbb{Q}}(f_m(d))$ is greater than 1. Then, one has:*

$$N_{\mathbb{Q}(f_m(d))/\mathbb{Q}}\big(f_m(d)\big) \in \bigcup_{a \in \{2,3,4,5,6,8,9,12,16\}} \pm a^{\downarrow} \quad =: N.$$

PROOF. By assumption, the numerator of $N_{\mathbb{Q}(f_m(d))/\mathbb{Q}}(f_m(d))$ has a prime factor $p \in \mathbb{P}$. Further, by Lemma 1.42, for every such prime factor $p \in \mathbb{P}$, one has

$$v_p\Big(N_{\mathbb{Q}(f_m(d))/\mathbb{Q}}(f_m(d))\Big) = 2\,v_p(f_m(d)) = 2\,v_p\left(\frac{(1-\zeta_m^{k_1})(1-\zeta_m^{k_2})}{(1-\zeta_m^{k_3})(1-\zeta_m^{k_4})}\right) \in \mathbb{N}. \tag{1.15}$$

Applying Equations (1.6) and (1.7), Proposition 1.31 and Lemma 1.43, one sees that $v_p(f_m(d))$ is a sum of at most four terms of the form $1/(p^{t'-1}(p-1))$ for various $t' \in \mathbb{N}$ with one or two positive terms and at most two negative ones. Let t be the smallest t' occurring in one of the positive terms. Then, Relation (1.15) particularly shows that

$$\frac{4}{p^{t-1}(p-1)} \geq 1$$

or, equivalently,

$$p^{t-1}(p-1) \leq 4. \tag{1.16}$$

One can see easily that the only possibilities for (1.16) are $p = 2$ and $t \in \{1,2,3\}$, or $p = 3$ and $t = 1$, or $p = 5$ and $t = 1$. Moreover, using (1.15) in conjunction with Equations (1.6) and (1.7) and Proposition 1.31, one can see the following. In the first case ($p = 2$ and $t \in \{1,2,3\}$), the only possibilities for powers of 2 dividing the numerator of $N_{\mathbb{Q}(f_m(d))/\mathbb{Q}}(f_m(d))$ are $2, 4, 8$ and 16, with room for a 3 as well in the cases 2 and 4, i.e., in the case where the 2-adic value of $N_{\mathbb{Q}(f_m(d))/\mathbb{Q}}(f_m(d))$ is 1 or 2. Hence, one obtains $N_{\mathbb{Q}(f_m(d))/\mathbb{Q}}(f_m(d)) \in \bigcup_{a\in\{2,4,6,8,12,16\}} \pm a^{\downarrow}$. In the second case ($p = 3$ and $t = 1$), the only possibilities for powers of 3 dividing the numerator of $N_{\mathbb{Q}(f_m(d))/\mathbb{Q}}(f_m(d))$ are 3 and 9, with room for a 2 or 4 as well in the case 3, i.e., in the case where the 3-adic value of $N_{\mathbb{Q}(f_m(d))/\mathbb{Q}}(f_m(d))$ is 1. Consequently, one obtains $N_{\mathbb{Q}(f_m(d))/\mathbb{Q}}(f_m(d)) \in \bigcup_{a\in\{3,6,9,12\}} \pm a^{\downarrow}$, whereas the last case ($p = 5$ and $t = 1$) immediately implies that $N_{\mathbb{Q}(f_m(d))/\mathbb{Q}}(f_m(d)) \in \pm 5^{\downarrow}$. □

LEMMA 1.47. *Let $m \geq 4$ and let $d := (k_1, k_2, k_3, k_4) \in D'_m$. Suppose that $f_m(d)$ is of degree 2 over $\mathbb{Q}$ and suppose that the absolute value of $N_{\mathbb{Q}(f_m(d))/\mathbb{Q}}(f_m(d))$ is smaller than 1. Then, one has:*

$$N_{\mathbb{Q}(f_m(d))/\mathbb{Q}}(f_m(d)) \in [N]^{-1},$$

with N as defined in Lemma 1.46.

PROOF. Note that, by Lemma 1.38, one has $f_m(d) > 0$, which implies that the absolute value of the norm $N_{\mathbb{Q}(f_m(d))/\mathbb{Q}}(f_m(d))$ is non-zero. The assertion now follows immediately from Lemma 1.44 and Lemma 1.46. □

THEOREM 1.48. *For any real quadratic algebraic number field $\mathbb{k}$, one has:*

$$N_{\mathbb{k}/\mathbb{Q}}\Big[\Big(\bigcup_{m\geq 4} f_m[D_m]\Big) \cap \mathbb{k}\Big] \subset \{-1,1\} \cup [N_1]^2 \cup N \cup [N]^{-1} =: N_2,$$

with N_1 as defined in Theorem 1.40 and N as defined in Lemma 1.46.

PROOF. Let $f_m(d) \in (\bigcup_{m\geq 4} f_m[D_m]) \cap \mathbb{k}$ for suitable $m \geq 4$ and $d \in D_m$. Note that, by Lemma 1.38, one has $f_m(d) > 1$ and hence $f_m(d) \neq 0$. First, suppose that $f_m(d)$ is of degree two over $\mathbb{Q}$, i.e., a primitive element of the field extension $\mathbb{k}/\mathbb{Q}$. Since $f_m(d) \neq 0$, the norm $N_{\mathbb{Q}(f_m(d))/\mathbb{Q}}(f_m(d)) = N_{\mathbb{k}/\mathbb{Q}}(f_m(d))$ is non-zero. Therefore, the absolute value of

$N_{\Bbbk/\mathbb{Q}}(f_m(d))$ is greater than zero. The following is divided into the three possible cases. First, assume that the absolute value of $N_{\Bbbk/\mathbb{Q}}(f_m(d))$ is greater than 1. Then, Lemma 1.46 immediately implies that $N_{\Bbbk/\mathbb{Q}}(f_m(d)) \in N$. Secondly, suppose that the absolute value of $N_{\Bbbk/\mathbb{Q}}(f_m(d))$ is smaller than 1. Then, Lemma 1.47 immediately implies that $N_{\Bbbk/\mathbb{Q}}(f_m(d)) \in [N]^{-1}$. Finally, if the absolute value of $N_{\Bbbk/\mathbb{Q}}(f_m(d))$ equals 1, one obviously gets

$$N_{\Bbbk/\mathbb{Q}}(f_m(d)) \in \{1, -1\}.$$

Now, suppose that $f_m(d) \in \mathbb{Q}$. Then, by Theorem 1.40, one has $f_m(d) \in N_1$ and hence $N_{\Bbbk/\mathbb{Q}}(f_m(d)) \in [N_1]^2$; cf. Remark 1.6. This completes the proof. □

1.1.5.2. *The general case.*

THEOREM 1.49. *For any $e \in \mathbb{N}$, there is a finite set $N_e \subset \mathbb{Q}$ such that, for all real algebraic number fields $\Bbbk$ having degree e, one has:*

$$N_{\Bbbk/\mathbb{Q}}\Big[\Big(\bigcup_{m\geq 4} f_m[D_m]\Big) \cap \Bbbk\Big] \subset N_e.$$

PROOF. It suffices to show the corresponding assertion for the superset D'_m of D_m. Let $e \in \mathbb{N}$ and let $\Bbbk$ be a real algebraic number fields $\Bbbk$ of degree e. Further, let $f_m(d) \in (\bigcup_{m\geq 4} f_m[D'_m]) \cap \Bbbk$ for suitable $m \geq 4$ and $d \in D_m$. By Lemma 1.13, and since one has

$$\mathbb{Q} \subset \mathbb{Q}\big(f_m(d)\big) \subset \Bbbk,$$

$f_m(d)$ is of degree $f := [\mathbb{Q}\big(f_m(d)\big) : \mathbb{Q}]$ over $\mathbb{Q}$, where f is a divisor of e. By Lemma 1.38, one has $f_m(d) \neq 0$. So, the norm $N_{\mathbb{Q}(f_m(d))/\mathbb{Q}}(f_m(d))$ is non-zero, hence its absolute value is greater than zero. Suppose that the absolute value of $N_{\mathbb{Q}(f_m(d))/\mathbb{Q}}(f_m(d))$ is greater than 1. Then, the numerator of $N_{\mathbb{Q}(f_m(d))/\mathbb{Q}}(f_m(d))$ has a prime factor, say $p \in \mathbb{P}$. Further, by Lemma 1.42, for every such prime factor $p \in \mathbb{P}$, one has

$$v_p\Big(N_{\mathbb{Q}(f_m(d))/\mathbb{Q}}\big(f_m(d)\big)\Big) = e\, v_p\big(f_m(d)\big) = e\, v_p\left(\frac{(1-\zeta_m^{k_1})(1-\zeta_m^{k_2})}{(1-\zeta_m^{k_3})(1-\zeta_m^{k_4})}\right) \in \mathbb{N}. \tag{1.17}$$

Applying Equations (1.6) and (1.7), Proposition 1.31 and Lemma 1.43, one sees that $v_p(f_m(d))$ is a sum of at most four terms of the form $1/(p^{t'-1}(p-1))$ for various $t' \in \mathbb{N}$ with one or two positive terms and at most two negative ones. Let t be the smallest t' occurring in one of the positive terms. Then, Relation (1.17) particularly shows that

$$\frac{2\,e}{p^{t-1}(p-1)} \geq 1$$

or, equivalently,

$$p^{t-1}(p-1) \leq 2\,e. \tag{1.18}$$

Similar to the argumentation in Lemma 1.46 above, one can now see that, by Relations (1.17) and (1.18) and the obvious fact that

$$p^{t-1}(p-1) \longrightarrow \infty$$

for fixed $p \in \mathbb{P}$ as $t \longrightarrow \infty$ (resp., for fixed $t \in \mathbb{N}$ as $\mathbb{P} \ni p \longrightarrow \infty$), there is a finite set, say N_f, such that $N_{\mathbb{Q}(f_m(d))/\mathbb{Q}}(f_m(d)) \in N_f$. Moreover, analogous to the argumentation in

Lemma 1.47 above, one can see that, if the absolute value of $N_{\mathbb{Q}(f_m(d))/\mathbb{Q}}(f_m(d))$ is smaller than one, then one has

$$N_{\mathbb{Q}(f_m(d))/\mathbb{Q}}(f_m(d)) \in [N_f]^{-1},$$

whereas the missing case, where the absolute value of $N_{\mathbb{Q}(f_m(d))/\mathbb{Q}}(f_m(d))$ equals 1, only gives

$$N_{\mathbb{Q}(f_m(d))/\mathbb{Q}}(f_m(d)) \in \{1,-1\}.$$

The transitivity of the norm (cf. Remark 1.6) now implies that $N_{\mathbb{k}/\mathbb{Q}}(f_m(d)) \in \big[\{1,-1\} \cup N_f \cup [N_f]^{-1}\big]^{\frac{e}{f}}$. Setting

$$N_e := \bigcup_{f \mid e} \big[\{1,-1\} \cup N_f \cup [N_f]^{-1}\big]^{\frac{e}{f}},$$

the assertion immediately follows, since the above analysis only depends on the fixed degree e. □

DEFINITION 1.50. We denote by $\mathbb{Q}_{>1}^{\mathbb{P}_{\leq 2}}$ the subset of $\mathbb{Q}$ consisting of all numbers greater than one, which have at most two prime factors in the numerator.

THEOREM 1.51. *For any real algebraic number field* $\mathbb{k}$, *one has:*

$$N_{\mathbb{k}/\mathbb{Q}}\Big[\big(\bigcup_{m \geq 4} f_m[D_m]\big) \cap \mathbb{k}\Big] \subset \pm\Big(\{1\} \cup \big[\mathbb{Q}_{>1}^{\mathbb{P}_{\leq 2}}\big]^{\pm 1}\Big).$$

PROOF. This follows from a careful analysis of the proof of Theorem 1.49. □

1.1.5.3. *Applications.* For our later application to discrete tomography of aperiodic model sets, we formulate some consequences. We are mainly interested in applications of the above results to real algebraic number fields of the form $\mathbb{k}_n$, in particular for $n \in \{5, 8, 10, 12\}$, in view of their practical relevance; see [**65**].

REMARK 1.52. Note that $\mathbb{k}_{10} = \mathbb{k}_5 = \mathbb{Q}(\sqrt{5})$, $\mathbb{k}_8 = \mathbb{Q}(\sqrt{2})$ and $\mathbb{k}_{12} = \mathbb{Q}(\sqrt{3})$. By Corollary 1.14, all of these fields are real, quadratic algebraic number fields.

COROLLARY 1.53. *For* $n \in \{5, 8, 10, 12\}$, *one has:*

$$N_{\mathbb{k}_n/\mathbb{Q}}\Big[\big(\bigcup_{m \geq 4} f_m[D_m]\big) \cap \mathbb{k}_n\Big] \subset N_2,$$

with N_2 *as defined in Theorem 1.48.*

PROOF. This is an immediate consequence of Theorem 1.48. □

DEFINITION 1.54. Let the map $\frac{\phi}{2}\colon \mathbb{N} \setminus \{1,2\} \longrightarrow \mathbb{N}$ be given by $n \longmapsto \phi(n)/2$.

REMARK 1.55. Recall that, for $n \in \mathbb{N} \setminus \{1,2\}$, $\frac{\phi}{2}(n)$ is the degree of $\mathbb{k}_n$ over $\mathbb{Q}$; cf. Corollary 1.14.

COROLLARY 1.56. *For all* $e \in \operatorname{Im}(\frac{\phi}{2})$, *there is a finite set* $N_e \subset \mathbb{Q}$ *such that, for all* $n \in (\frac{\phi}{2})^{-1}[\{e\}]$, *one has:*

$$N_{\mathbb{k}_n/\mathbb{Q}}\Big[\big(\bigcup_{m \geq 4} f_m[D_m]\big) \cap \mathbb{k}_n\Big] \subset N_e.$$

PROOF. This is an immediate consequence of Theorem 1.49. □

1.2. Delone sets, Meyer sets, and model sets

1.2.1. Delone sets and Meyer sets.

DEFINITION 1.57. Consider a set $\Lambda \subset \mathbb{R}^d$ with $d \in \mathbb{N}$.

(a) Λ is called *uniformly discrete* if there is a radius $r > 0$ such that every ball $B_r(x)$ with $x \in \mathbb{R}^d$ contains at most one point of Λ.
(b) Λ is called *relatively dense* if there is a radius $R > 0$ such that every ball $B_R(x)$ with $x \in \mathbb{R}^d$ contains at least one point of Λ.
(c) Λ is called a *Delone set* (or *Delaunay set*) if it is both uniformly discrete and relatively dense.
(d) Λ has *finite local complexity* if $\Lambda - \Lambda$ is discrete and closed.
(e) A Delone set Λ is a *Meyer set* if $\Lambda - \Lambda$ is uniformly discrete.
(f) Λ is called *aperiodic* if it has no translational symmetries, i.e., if

$$\left\{ t \in \mathbb{R}^d \,\middle|\, t + \Lambda = \Lambda \right\} = \{0\} .$$

(g) Λ is *repetitive* if, given any set of the form $\Lambda \cap B_r(x)$ (called *patch of diameter* r), where $x \in \mathbb{R}^d$, there is a radius $R > 0$ such that any ball $B_R(y)$, where $y \in \mathbb{R}^d$, contains at least one translate of this patch.
(h) Λ has *frequencies of repetition of finite patches* if, for every finite patch, the number of occurrences of translates of this patch per unit volume in the ball $B_r(0)$ of radius $r > 0$ about the origin 0 approaches a non-negative limit as $r \to \infty$.

REMARK 1.58. It follows immediately from the definition that every Meyer set Λ has finite local complexity. Note further that a subset Λ of $\mathbb{R}^d$, $d \in \mathbb{N}$, has finite local complexity if and only if for every $r > 0$ there are, up to translation, only finitely many point sets (called *patches of diameter* r) of the form $\Lambda \cap B_r(x)$, where $x \in \mathbb{R}^d$; cf. [**60**, Proposition 2.3].

LEMMA 1.59. *Let $d \geq 2$ and let $\Lambda \subset \mathbb{R}^d$ be relatively dense. Then, the set of Λ-directions is dense in $\mathbb{S}^{d-1}$.*

PROOF. We may assume, without loss of generality, that $0 \in \Lambda$. Let $u \in \mathbb{S}^{d-1}$ and let $B_\varepsilon(u) \cap \mathbb{S}^{d-1}$ be an arbitrary open ε-neighbourhood of u in $\mathbb{S}^{d-1}$. Without restriction, let $\varepsilon < 1$. Then, $B_\varepsilon(u) \cap \mathbb{S}^{d-1}$ is a $(d-1)$-dimensional open ball, i.e., homeomorphic to the open ball $B_1(0) \subset \mathbb{R}^{d-1}$. Consider the smallest convex cone C in $\mathbb{R}^d$ with apex 0 and containing the set $B_\varepsilon(u) \cap \mathbb{S}^{d-1}$, i.e.,

$$C := \left\{ \sum_{j=1}^{n} \lambda_j x_j \,\middle|\, \begin{array}{l} n \in \mathbb{N},\, x_1, \dots, x_n \in B_\varepsilon(u) \cap \mathbb{S}^{d-1}, \\ \mathbb{R} \ni \lambda_1, \dots, \lambda_n \geq 0 \end{array} \right\} .$$

Since Λ is relatively dense, there is a radius $R > 0$ such that every open ball $B_R(z)$ with $z \in \mathbb{R}^d$ contains at least one element of Λ. Clearly, the interior $\operatorname{int}(C) = C \setminus \{0\}$ of the convex cone C contains open balls of arbitrary large radius, hence points λ of Λ; see Figure 1.2 for an illustration. This completes the proof, since any $\lambda \in \Lambda \cap \operatorname{int}(C)$ yields the Λ-direction $u_\lambda \in B_\varepsilon(u) \cap \mathbb{S}^{d-1}$. □

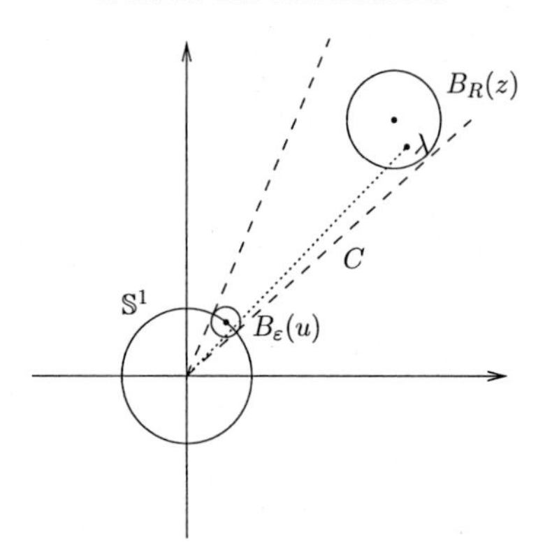

FIGURE 1.2. Illustration of the proof of Lemma 1.59.

1.2.2. Algebraic Delone sets.

DEFINITION 1.60. For $\Lambda \subset \mathbb{R}^2 \cong \mathbb{C}$, we set

$$\mathbb{K}_\Lambda := \mathbb{Q}\left((\Lambda - \Lambda) \cup (\overline{\Lambda - \Lambda})\right) .$$

and, further, $\Bbbk_\Lambda := \mathbb{K}_\Lambda \cap \mathbb{R}$.

The class of *algebraic Delone sets* is defined as follows.

DEFINITION 1.61. A Delone set $\Lambda \subset \mathbb{R}^2$ is called an *algebraic Delone set* if it satisfies the following properties:

(A) $[\mathbb{K}_\Lambda : \mathbb{Q}] < \infty$.
(B) For all finite subsets F of $\mathbb{K}_\Lambda$, there is a homothety $h : \mathbb{R}^2 \longrightarrow \mathbb{R}^2$ such that $h[F] \subset \Lambda$.

The next result shows that the property of being an algebraic Delone set is invariant under translations.

LEMMA 1.62. *Let $\Lambda \subset \mathbb{R}^2$. The following statements are equivalent:*

(i) *Λ is an algebraic Delone set.*
(ii) *For all $t \in \mathbb{R}^2$, $t + \Lambda$ is an algebraic Delone set.*

PROOF. Direction (ii) $\Rightarrow$ (i) is trivial. In order to prove direction (i) $\Rightarrow$ (ii), let Λ be an algebraic Delone set and consider the translate $t + \Lambda$, where $t \in \mathbb{R}^2$. First, note that $t + \Lambda$ is again a Delone set in $\mathbb{R}^2$. In order to prove property (A) for $t + \Lambda$, note that $\mathbb{K}_{t+\Lambda} = \mathbb{K}_\Lambda$. Since Λ has property (A), it follows that

$$[\mathbb{K}_{t+\Lambda} : \mathbb{Q}] = [\mathbb{K}_\Lambda : \mathbb{Q}] < \infty .$$

For property (B), let F be a finite subset of $\mathbb{K}_{t+\Lambda} = \mathbb{K}_\Lambda$. By property (B) for Λ, there is a homothety $h : \mathbb{R}^2 \longrightarrow \mathbb{R}^2$ such that $h[F] \subset \Lambda$. Hence, $(h + t)[F] \subset t + \Lambda$, where $h + t \colon \mathbb{R}^2 \longrightarrow \mathbb{R}^2$ is the homothety given by $(h + t)(z) := h(z) + t$. $\square$

REMARK 1.63. It will turn out below that the class of *cyclotomic model sets* is contained in the class of algebraic Delone sets (Lemma 1.83). Further, it will be shown in Section 2.3 that an important uniqueness problem in discrete tomography of algebraic Delone sets, i.e., the problem of determining the set of convex subsets of algebraic Delone sets by a small number of X-rays, allows a deep analysis.

LEMMA 1.64. *For an algebraic Delone set Λ, the field $\mathbb{k}_\Lambda$ is a real algebraic number field.*

PROOF. By definition, $\mathbb{k}_\Lambda$ is the maximal real subfield of $\mathbb{K}_\Lambda$ and, by property (A) in the definition of algebraic Delone sets, $\mathbb{k}_\Lambda$ is indeed a finite extension of $\mathbb{Q}$. □

Lagarias [**45**] defined the notion of *finitely generated Delone sets* $\Lambda \subset \mathbb{R}^d$, where $d \in \mathbb{N}$. These are Delone sets Λ having the property that the Abelian group $\langle \Lambda - \Lambda \rangle_{\mathbb{Z}}$ generated by the difference set $\Lambda - \Lambda$ of Λ is finitely generated. The latter is the case if and only if the Abelian group $\langle \Lambda \rangle_{\mathbb{Z}}$ generated by Λ itself is finitely generated. Note that, for a Delone set Λ, the property that the the Abelian group $\langle \Lambda \rangle_{\mathbb{Z}}$ is finitely generated can be seen as a weak form of long-range order of Λ under translations. The last property is always fulfilled by Delone sets of finite local complexity, which are also called *Delone sets of finite type*; see [**45**, Theorem 2.1]. Note that the finitely generated subgroups of $\mathbb{R}^d$ are exactly the free subgroups of $\mathbb{R}^d$ of finite rank. There is the following sufficient condition for an algebraic Delone set to be finitely generated.

PROPOSITION 1.65. *Let $\Lambda \subset \mathbb{R}^2 \cong \mathbb{C}$ be an algebraic Delone set. If $\Lambda - \Lambda$ is contained in the ring of algebraic integers, then Λ is a finitely generated Delone set.*

PROOF. If $\Lambda - \Lambda$ only consists of algebraic integers, then the Abelian group $\langle \Lambda - \Lambda \rangle_{\mathbb{Z}}$ obviously is contained in the ring of integers of the algebraic number field $\mathbb{K}_\Lambda$. In particular, the ring of integers of $\mathbb{K}_\Lambda$ is a free Abelian group of rank $[\mathbb{K}_\Lambda : \mathbb{Q}] < \infty$. It follows that its subgroup $\langle \Lambda - \Lambda \rangle_{\mathbb{Z}}$ is finitely generated (in fact, it is free of rank r, where $r \leq [\mathbb{K}_\Lambda : \mathbb{Q}]$); see [**18**, Chapter 2, Section 2]. □

1.2.3. Model sets.

1.2.3.1. *General setting.* By definition, *model sets* arise from so-called *cut and project schemes*. These are commutative diagrams of the following form; compare [**48**] and see [**7**] for a gentle introduction with many illustrations.

$$
\begin{array}{ccccc}
\mathbb{R}^d & \overset{\pi}{\longleftarrow} & \mathbb{R}^d \times H & \overset{\pi_{\text{int}}}{\longrightarrow} & H \\
\cup & & \cup \ \text{lattice} & & \cup \ \text{dense} \\
\pi[\widetilde{L}] & \overset{1-1}{\longleftrightarrow} & \widetilde{L} & \longrightarrow & \pi_{\text{int}}[\widetilde{L}]
\end{array}
\tag{1.19}
$$

Here, H is some locally compact Abelian group, π and π_{int} are the canonical projections, and $\widetilde{L}$ is a lattice in $\mathbb{R}^d \times H$, i.e., $\widetilde{L}$ is a discrete subgroup of $\mathbb{R}^d \times H$ such that the quotient group

$$(\mathbb{R}^d \times H) \,/\, \widetilde{L}$$

is compact. Further, $\pi_{\text{int}}[\widetilde{L}]$ is a dense subset of H and the restriction of π to $\widetilde{L}$ is assumed to be injective. Writing $L := \pi[\widetilde{L}]$, one can define a map $.^\star\colon L \longrightarrow H$ by $x \longmapsto \pi_{\text{int}}(\pi|_{\widetilde{L}}^{-1}(x))$. Then, one has $[L]^\star = \pi_{\text{int}}[\widetilde{L}]$. If the map $.^\star$ is injective, we denote the inverse of its co-restriction $.^\star\colon L \longrightarrow [L]^\star$ by $.^{-\star}\colon [L]^\star \longrightarrow L$.

DEFINITION 1.66. (a) Given the cut and project scheme (1.19), a subset $W \subset H$ is called a *window* if $\varnothing \neq \text{int}(W) \subset W \subset \text{cl}(\text{int}(W))$ and $\text{cl}(\text{int}(W))$ is compact.

(b) Given any window $W \subset H$, and any $t \in \mathbb{R}^d$, we obtain a *model set*

$$\varLambda(t, W) := t + \varLambda(W)$$

relative to the cut and project scheme by setting

$$\varLambda(W) := \{x \in L \,|\, x^\star \in W\}\,.$$

Further, $\mathbb{R}^d$ (resp., H) is called the *physical* (resp., *internal*) space and W is also referred to as the *window* of $\varLambda(t, W)$. The map $.^\star\colon L \longrightarrow H$, as defined above, is the so-called *star map*. Finally, L is called the *underlying* $\mathbb{Z}$*-module* of $\varLambda(t, W)$.

REMARK 1.67. Note that L is indeed a $\mathbb{Z}$-module. Further, the star map is a homomorphism of Abelian groups.

For details about model sets and general background material, see [**48**] and [**14**]; see [**7**] for detailed graphical illustrations of the projection method. Note that, if the internal space is Euclidean, i.e., $H = \mathbb{R}^k$ for some $k \in \mathbb{N}$, then the star map $.^\star\colon L \longrightarrow H$ can be uniquely extended to a $\mathbb{Q}$-linear map of the $\mathbb{Q}$-span $\mathbb{Q}L$ of the $\mathbb{Z}$-module L, but *not* to an $\mathbb{R}$-linear map of all of $\mathbb{R}^d$.

REMARK 1.68. The translation vector t in Definition 1.66 stresses an intrinsic character of model sets. While the structure model specifies the cut and project scheme k, H and $\widetilde{L}$, and also the window W, a natural choice of the origin is usually not possible.

REMARK 1.69. Without loss of generality, we may assume that the stabilizer H_W of the window W, i.e.,

$$H_W := \{h \in H \,|\, h + W = W\},$$

is the trivial subgroup of H, i.e., $H_W = \{0\}$; see [**13, 59, 60**]. Observe that the latter is always the case if H is some Euclidean space, i.e., if one has $H = \mathbb{R}^k$ for some suitable $k \in \mathbb{N}$.

DEFINITION 1.70. Let $\varLambda := \varLambda(t, W) \subset \mathbb{R}^d$ be a model set as defined in Definition 1.66.

(a) $\varLambda$ is called *regular* if the boundary $\text{bd}(W)$ of the window W has (Haar) measure 0 in H.

(b) $\varLambda$ is called *generic* if $[L]^\star \cap \text{bd}(W) = \varnothing$.

The following remark collects some properties of model sets; for details see [**48**].

REMARK 1.71. The model set $\varLambda := \varLambda(t, W) \subset \mathbb{R}^d$ is an aperiodic Meyer set and thus has finite local complexity; see [**48**]. Note that the aperiodicity is equivalent to the injectivity of the star map. In fact, the kernel of the star map is the group of translational symmetries of $\varLambda$; see [48] again. If $\varLambda$ is regular, then $\varLambda$ is *pure point diffractive*; cf. [**60**]. If $\varLambda$ is generic, then $\varLambda$ is *repetitive*; see [**60**]. If $\varLambda$ is regular, then $\varLambda$ has frequencies of repetition of finite patches; cf. [**59**]. Although not all of these properties are used below, they actually show the similarity of aperiodic model sets with lattices – except their lack of periods.

1.2.3.2. *Cyclotomic model sets.* Before we can present the cut and project scheme from which *cyclotomic* model sets arise, we need the following remark. In the following let $n \in \mathbb{N} \setminus \{1,2\}$ (this means that $\phi(n) > 1$).

REMARK 1.72. The elements of the Galois group $G(\mathbb{K}_n/\mathbb{Q})$ come in pairs of complex conjugate automorphisms. Let the set $\{\sigma_1, \dots, \sigma_{\phi(n)/2}\}$ arise from $G(\mathbb{K}_n/\mathbb{Q})$ by choosing exactly one automorphism from each such pair, where we assume that σ_1 is either the identity or the complex conjugation. Every such choice induces a map

$$.^{\sim_n} : \mathcal{O}_n \longrightarrow \mathbb{R}^2 \times (\mathbb{R}^2)^{\frac{\phi(n)}{2}-1},$$

given by

$$z \longmapsto \Big(\sigma_1(z), \big(\sigma_2(z), \dots, \sigma_{\frac{\phi(n)}{2}}(z)\big)\Big).$$

With the understanding that for $\phi(n) = 2$ (this means $n \in \{3,4,6\}$), the singleton

$$(\mathbb{R}^2)^{\phi(n)/2-1} = (\mathbb{R}^2)^0$$

is the trivial (locally compact) Abelian group $\{0\}$, each such choice induces, via projection on the second factor, a so-called *star map*

$$.^{\star_n} : \mathcal{O}_n \longrightarrow (\mathbb{R}^2)^{\frac{\phi(n)}{2}-1},$$

i.e., a map given by $.^{\star_n} \equiv 0$, if $n \in \{3,4,6\}$, and by $z \longmapsto \big(\sigma_2(z), \dots, \sigma_{\phi(n)/2}(z)\big)$ otherwise. Then, the map $.^{\sim_n}$ is called a *Minkowski embedding* of $\mathcal{O}_n$ and $[\mathcal{O}_n]^{\sim_n}$ is called a *Minkowski representation* of the maximal order $\mathcal{O}_n$ of $\mathbb{K}_n$; cf. Proposition 1.23(a) and see [18, Ch. 2, Sec. 3]. It follows that $[\mathcal{O}_n]^{\sim_n}$ is a (full) lattice in $\mathbb{R}^2 \times (\mathbb{R}^2)^{\phi(n)/2-1}$, meaning that it is a co-compact discrete subgroup of the Abelian group $\mathbb{R}^2 \times (\mathbb{R}^2)^{\phi(n)/2-1}$ (the quotient group is a $\phi(n)$-dimensional torus). Here, this is equivalent to the existence of $\phi(n)$ $\mathbb{R}$-linearly independent vectors in $\mathbb{R}^2 \times (\mathbb{R}^2)^{\phi(n)/2-1}$ whose $\mathbb{Z}$-linear hull equals $[\mathcal{O}_n]^{\sim_n}$, compare [**18**, Ch. 2, Sec. 3 and 4]. In fact, the set

$$\Big\{1^{\sim_n}, (\zeta_n)^{\sim_n}, \dots, (\zeta_n^{\phi(n)-1})^{\sim_n}\Big\}$$

has this property; cf. Proposition 1.23 and Remark 1.24. Further, note that the image $[\mathcal{O}_n]^{\star_n}$ is dense in $(\mathbb{R}^2)^{\phi(n)/2-1}$. This follows for instance from the existence of a PV-number λ of (full) degree $\phi(n)/2$ in $\mathcal{O}_n$; see Lemma 1.29. In fact, if $\lambda \in \mathcal{O}_n$, one can see that, for every $k \in \mathbb{N}$, the subset

$$\begin{aligned} &\Big\{(\lambda^k z)^{\star_n} \,\Big|\, z \in \{1, \zeta_n, \dots, \zeta_n^{\phi(n)-1}\}\Big\} \\ = \; &\Big\{\Big(\big(\sigma_2(\lambda)\big)^k \sigma_2(z), \dots, \big(\sigma_{\frac{\phi(n)}{2}}(\lambda)\big)^k \sigma_{\frac{\phi(n)}{2}}(z)\Big) \,\Big|\, z \in \{1, \zeta_n, \dots, \zeta_n^{\phi(n)-1}\}\Big\} \end{aligned}$$

of $[\mathcal{O}_n]^{\star_n}$ is a generating set of the vector space $(\mathbb{R}^2)^{\phi(n)/2-1}$ over $\mathbb{R}$. If, in addition, λ is a PV-number of (full) degree $\phi(n)/2$, then the set $\{\sigma_1(\lambda), \sigma_2(\lambda), \dots, \sigma_{\phi(n)/2}(\lambda)\}$ equals the set of (algebraic) conjugates of λ and, moreover, one has the inclusion $\{|\sigma_2(\lambda)|, \dots, |\sigma_{\phi(n)/2}(\lambda)|\} \subset (0,1)$; see the proof of Lemma 1.80 below. It follows that, for every $\varepsilon > 0$, there is an $\mathbb{R}$-basis $B_\varepsilon \subset [\mathcal{O}_n]^{\star_n}$ of $(\mathbb{R}^2)^{\phi(n)/2-1}$ such that all elements b of B_ε satisfy $\|b\| \leq \varepsilon$. Clearly, this implies the denseness of $[\mathcal{O}_n]^{\star_n}$ in $(\mathbb{R}^2)^{\phi(n)/2-1}$.

The class of *cyclotomic* model sets arises from cut and project schemes of the following form. Consider the following diagram (cut and project scheme), where we follow Moody [**48**], modified in the spirit of the algebraic setting of Pleasants [**51**].

$$\begin{array}{ccccc} \mathbb{R}^2 & \xleftarrow{\ \pi\ } & \mathbb{R}^2 \times (\mathbb{R}^2)^{\frac{\phi(n)}{2}-1} & \xrightarrow{\ \pi_{\text{int}}\ } & (\mathbb{R}^2)^{\frac{\phi(n)}{2}-1} \\ \cup & & \cup \text{ lattice} & & \cup \text{ dense} \\ \mathcal{O}_n & \xleftrightarrow{\ 1\text{–}1\ } & [\mathcal{O}_n]^{\sim n} & \longrightarrow & [\mathcal{O}_n]^{\star n} \end{array} \tag{1.20}$$

Here, we always choose σ_1 as the identity rather than the complex conjugation; see Remark 1.72. Hence, one has

$$[\mathcal{O}_n]^{\sim n} = \Big\{ \Big(z, \underbrace{(\sigma_2(z), \dots, \sigma_{\frac{\phi(n)}{2}}(z))}_{=z^{\star n}}\Big) \,\Big|\, z \in \mathcal{O}_n \Big\}.$$

DEFINITION 1.73. Let $n \in \mathbb{N} \setminus \{1,2\}$. Given any window $W \subset (\mathbb{R}^2)^{\phi(n)/2-1}$ and any $t \in \mathbb{R}^2$, we obtain a planar model set

$$\varLambda_n(t, W) := t + \varLambda_n(W)$$

relative to any choice of the set $\{\sigma_j \,|\, j \in \{2, \dots, \phi(n)/2\}\}$ as described above by setting

$$\varLambda_n(W) := \{z \in \mathcal{O}_n \,|\, z^{\star n} \in W\}.$$

The dimension c of the internal space of $\varLambda_n(t,W) \in \mathcal{M}(\mathcal{O}_n)$, i.e., $c = \phi(n) - 2$, is called the *co-dimension* of $\varLambda_n(t,W)$. We set

$$\mathcal{M}(\mathcal{O}_n) := \Big\{ \varLambda_n(t,W) \,\Big|\, t \in \mathbb{R}^2, W \subset (\mathbb{R}^2)^{\frac{\phi(n)}{2}-1} \text{ is a window} \Big\}.$$

Then, the class $\mathcal{CM}$ of *cyclotomic* model sets is defined as

$$\mathcal{CM} := \bigcup_{n \in \mathbb{N}\setminus\{1,2\}} \mathcal{M}(\mathcal{O}_n).$$

Let $W \subset (\mathbb{R}^2)^{\phi(n)/2-1}$ be a window and let a star map $.^{\star n}$ be given (as described above). We let $\mathcal{M}_g(\mathcal{O}_n)$ be the set of generic elements of $\mathcal{M}(\mathcal{O}_n)$ and define

$$\mathcal{M}_g^{\star n}(W) := \left\{ \varLambda_n^{\star n}(t, \tau + W) \,\middle|\, \begin{array}{l} t \in \mathbb{R}^2, \tau \in (\mathbb{R}^2)^{\frac{\phi(n)}{2}-1} \text{ and} \\ [\mathcal{O}_n]^{\star n} \cap \mathrm{bd}(\tau + W) = \varnothing \end{array} \right\} \subset \mathcal{M}_g(\mathcal{O}_n),$$

where the elements $\varLambda_n^{\star n}(t, \tau + W)$ of this set are understood to be defined by use of the fixed star map $.^{\star n}$, i.e.,

$$\varLambda_n^{\star n}(t, \tau + W) = t + \{z \in \mathcal{O}_n \,|\, z^{\star n} \in \tau + W\}.$$

The elements of the set $\mathcal{M}_g^{\star n}(W)$ are called $\mathcal{M}_g^{\star n}(W)$*-sets.*

REMARK 1.74. Let $n \in \mathbb{N} \setminus \{1,2\}$. Note that, for $n = 4$ (resp., $n \in \{3,6\}$), the set $\mathcal{M}_g^{\star n}(W)$ simply consists of all translates of the square lattice $\mathcal{O}_4$ (resp., triangular lattice $\mathcal{O}_3 = \mathcal{O}_6$).

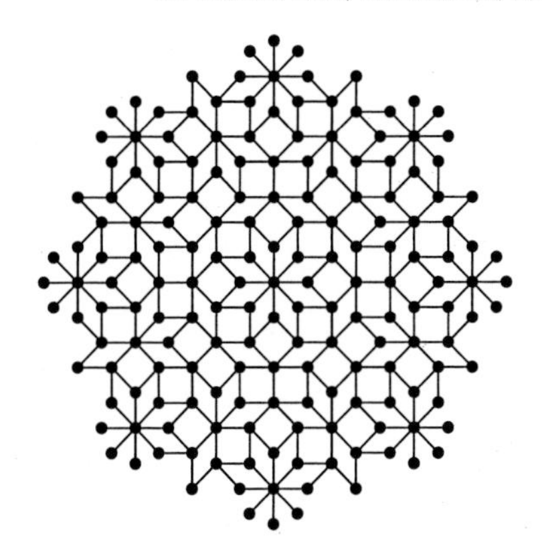

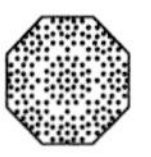

FIGURE 1.3. A central patch of the eightfold Ammann-Beenker tiling with vertex set $\varLambda_{\mathrm{AB}}$ (left) and the $.^{\star_8}$-image of $\varLambda_{\mathrm{AB}}$ inside the octagonal window in the so-called internal space (right), with relative scale as described in the text.

REMARK 1.75. We refer the reader to [**48, 51**] for details and related general settings. Note that the co-dimension of a cyclotomic model set is always an even number. Moreover, it is zero if and only if $n \in \{3,4,6\}$. Setting $\varLambda := \varLambda_n(t, W) \subset \mathbb{R}^2$, one has that $\varLambda$ is an aperiodic cyclotomic model set if and only if $n \notin \{3,4,6\}$, i.e., the translates of the square (resp., triangular) lattice are the only cyclotomic model sets with translation symmetries; compare Remark 1.71. Moreover, if $\varLambda$, for a given n, is both generic and regular, and, if the window W has m-fold cyclic symmetry with m a divisor of $N(n)$ ($N(n)$ is the function from (1.2)) and all in a suitable representation of the cyclic group C_m of order m, then $\varLambda$ has m-fold cyclic symmetry in the sense of symmetries of LI-classes, meaning that a discrete structure has a certain symmetry if the original and the transformed structure are locally indistinguishable (LI); see [**5**] for details.

EXAMPLE 1.76. The first two examples of cyclotomic model sets are elements of $\mathcal{M}(\mathcal{O}_n)$ of the form $\varLambda_n(0, W)$ for $n \in \{3,4\}$ and hence, necessarily, $W = \{0\}$.

(SQ) The planar generic regular periodic cyclotomic model set with 4-fold cyclic symmetry associated with the well-known square tiling is the square lattice, which can be described in algebraic terms as $\varLambda_{\mathrm{SQ}} := \varLambda_4(0, W) = \mathbb{Z}[i] = \mathcal{O}_4$; see Figure 1.1.

(TRI) The planar generic regular periodic cyclotomic model set with 6-fold cyclic symmetry associated with the well-known triangle tiling is the triangle lattice (also commonly known as the hexagonal lattice), which can be described in algebraic terms as $\varLambda_{\mathrm{TRI}} := \varLambda_3(0, W) = \mathcal{O}_3$; see Figure 1.1.

All further examples below are aperiodic cyclotomic model sets of the form $\varLambda_n(0, W) \in \mathcal{M}(\mathcal{O}_n)$ for suitable $n \in \mathbb{S}$ and satisfy $0 \in \mathrm{int}(W)$.

(AB) The planar generic regular aperiodic cyclotomic model set with 8-fold cyclic symmetry associated with the well-known Ammann-Beenker tiling can be described in

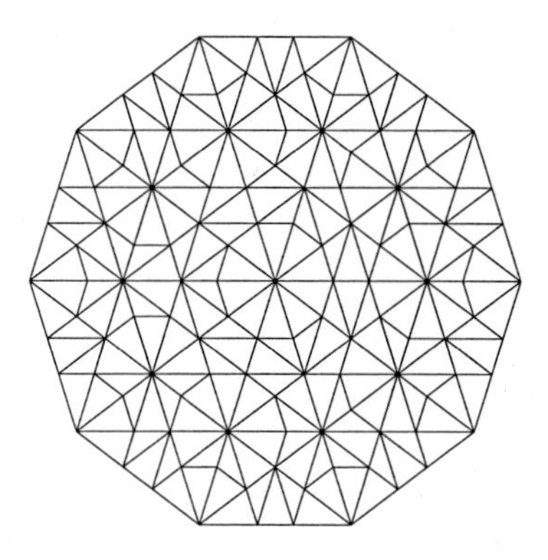

FIGURE 1.4. A central patch of the tenfold symmetric Tübingen triangle tiling.

algebraic terms as

$$\varLambda_{\mathrm{AB}} := \{z \in \mathcal{O}_8 \mid z^{\star_8} \in O\},$$

where the star map $.^{\star_8}$ is the Galois automorphism in $G(\mathbb{K}_8/\mathbb{Q})$, defined by $\zeta_8 \longmapsto \zeta_8^3$ (cf. Proposition 1.11), and the window O is the regular octagon centred at the origin, with vertices in the directions that arise from the 8th roots of unity by a rotation through $\pi/8$, and of unit edge length; see [**2, 10, 37**]. This construction also gives a tiling with squares and rhombi, both having edge length 1; see Figure 1.3.

(TTT) The planar regular aperiodic cyclotomic model set with 10-fold cyclic symmetry associated with the Tübingen triangle tiling [**11, 12**] can be described in algebraic terms as

$$\varLambda_{\mathrm{TTT}}^{u} := \{z \in \mathcal{O}_5 \mid z^{\star_5} \in u + W\},$$

where the star map $.^{\star_5}$ is the Galois automorphism in $G(\mathbb{K}_5/\mathbb{Q})$, defined by $\zeta_5 \longmapsto \zeta_5^2$ (cf. Proposition 1.11), and the window W is the regular decagon centred at the origin, with vertices in the directions that arise from the 10th roots of unity by a rotation through $\pi/10$, and of edge length $\tau/\sqrt{\tau+2}$, where τ denotes the golden ratio, i.e., $\tau = (1+\sqrt{5})/2$. Furthermore, u is an element of $\mathbb{R}^2$. $\varLambda_{\mathrm{TTT}}^{0}$ is not generic, while generic examples are obtained by shifting the window, i.e., $\varLambda_{\mathrm{TTT}}^{u}$ is generic for almost all $u \in \mathbb{R}^2$. Generic $\varLambda_{\mathrm{TTT}}^{u}$ always give a triangle tiling with long (short) edges of length 1 $(1/\tau)$. See Figure 1.4 for a generic example which we call $\varLambda_{\mathrm{TTT}}$; different generic choices of u result in locally indistinguishable Tübingen triangle tilings.

(S) The planar regular aperiodic cyclotomic model set with 12-fold cyclic symmetry associated with the shield tiling [**37**] can be described in algebraic terms as

$$\varLambda_{\mathrm{S}}^{u} := \{z \in \mathcal{O}_{12} \mid z^{\star_{12}} \in u + W\},$$

where the star map $.^{\star_{12}}$ is the Galois automorphism in $G(\mathbb{K}_{12}/\mathbb{Q})$, defined by $\zeta_{12} \longmapsto \zeta_{12}^5$ (cf. Proposition 1.11), and the window W is the regular dodecagon centred at the origin, with vertices in the directions that arise from the 12th roots of unity by

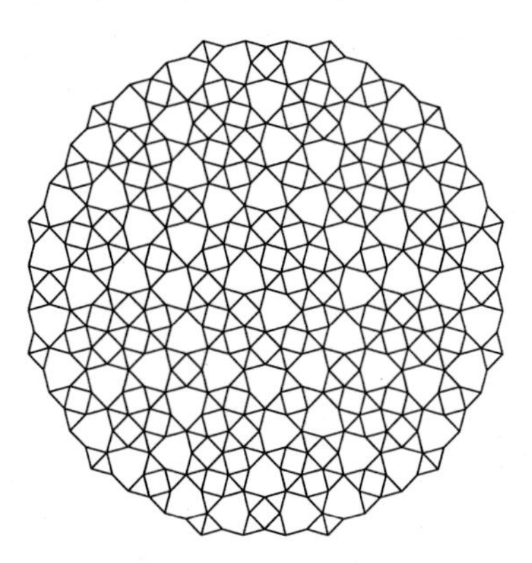

FIGURE 1.5. A central patch of the twelvefold symmetric shield tiling.

a rotation through $\pi/12$, and of edge length 1. Again, u is an element of $\mathbb{R}^2$. Λ_S^0 is not generic, while generic examples are obtained by shifting the window, i.e., Λ_S^u is generic for almost all $u \in \mathbb{R}^2$. The shortest distance between points in a generic Λ_S^u is $(\sqrt{3}-1)/\sqrt{2}$. Joining such points by edges results in a shield tiling, i.e., a tiling with triangles, squares and so-called shields as tiles, all having edge length $(\sqrt{3}-1)/\sqrt{2}$. See Figure 1.5 for a generic example which we call Λ_S; different generic choices of u result in locally indistinguishable shield tilings.

DEFINITION 1.77. Let $n \in \mathbb{N} \setminus \{1,2,3,4,6\}$, let $\lambda \in \mathcal{O}_n$, and let $\mathcal{O}_n \longrightarrow [\mathcal{O}_n]^{\sim_n}$ be a fixed Minkowski embedding of $\mathcal{O}_n$. Further, let $.^{\star_n}$ be the induced star map; see Remark 1.72.

(a) $m_\lambda^{(n)}$ denotes the $\mathbb{Z}$-module endomorphism of $\mathcal{O}_n$, given by $z \longmapsto \lambda z$.

(b) $(m_\lambda^{(n)})^{\sim_n}$ denotes the $\mathbb{Z}$-module (lattice) endomorphism of $[\mathcal{O}_n]^{\sim_n}$, given by

$$(z, z^{\star_n}) \longmapsto (\lambda z, (\lambda z)^{\star_n}).$$

(c) $(m_\lambda^{(n)})^{\star_n}$ denotes the $\mathbb{Z}$-module endomorphism of $[\mathcal{O}_n]^{\star_n}$, given by

$$z^{\star_n} \longmapsto (\lambda z)^{\star_n}.$$

REMARK 1.78. Note that, for $n \notin \{3,4,6\}$, every star map $.^{\star_n}$ is a $\mathbb{Z}$-module monomorphism; cf. Remark 1.75. Let $n \in \mathbb{N} \setminus \{1,2,3,4,6\}$, let $\lambda \in \mathcal{O}_n$ (resp., $\lambda \in \mathcal{O}_n^\times$), and let $.^{\sim_n} \colon \mathcal{O}_n \longrightarrow [\mathcal{O}_n]^{\sim_n}$ be a fixed Minkowski embedding of $\mathcal{O}_n$. Further, let $.^{\star_n}$ be the induced star map. Then, the $\mathbb{Z}$-module endomorphism (resp., automorphism) $m_\lambda^{(n)}$ of $\mathcal{O}_n$ corresponds via the chosen Minkowski embedding of $\mathcal{O}_n$ to the $\mathbb{Z}$-module (lattice) endomorphism (resp., automorphism) $(m_\lambda^{(n)})^{\sim_n}$ of $[\mathcal{O}_n]^{\sim_n}$. Further, $m_\lambda^{(n)}$ corresponds via the $\mathbb{Z}$-module automorphism $.^{\star_n} \colon \mathcal{O}_n \longrightarrow [\mathcal{O}_n]^{\star_n}$, given by the star map, to the $\mathbb{Z}$-module endomorphism (resp., automorphism) $(m_\lambda^{(n)})^{\star_n}$ of $[\mathcal{O}_n]^{\star_n}$.

DEFINITION 1.79. For $n \in \mathbb{N} \setminus \{1,2,3,4,6\}$, we denote by $\|\cdot\|_\infty$ the maximum norm on $(\mathbb{R}^2)^{\phi(n)/2-1}$ with respect to the Euclidean norm on all factors $\mathbb{R}^2$.

LEMMA 1.80. *Let $n \in \mathbb{N} \setminus \{1,2,3,4,6\}$, let $.^{\sim_n} \colon \mathcal{O}_n \longrightarrow [\mathcal{O}_n]^{\sim_n}$ be a fixed Minkowski embedding of $\mathcal{O}_n$, and let $.^{\star_n}$ be the induced star map. Then, for any PV-number λ of (full) degree $\phi(n)/2$ in $\mathcal{o}_n$, the $\mathbb{Z}$-module endomorphism $(m_\lambda^{(n)})^{\star_n}$ is contractive, i.e., there is a $\xi \in (0,1)$ such that the inequality*

$$\|(m_\lambda^{(n)})^{\star_n}(z^{\star_n})\|_\infty \leq \xi \, \|z^{\star_n}\|_\infty$$

holds for all $z \in \mathcal{O}_n$.

PROOF. Since λ is an algebraic integer of full degree $\phi(n)/2$, the set

$$\{\sigma_1(\lambda), \dots, \sigma_{\phi(n)/2}(\lambda)\}$$

equals the set of (algebraic) conjugates of λ. To see this, note that the set of (co-)restrictions

$$\left\{\sigma_1|_{\Bbbk_n}^{\Bbbk_n}, \dots, \sigma_{\phi(n)/2}|_{\Bbbk_n}^{\Bbbk_n}\right\}$$

equals the Galois group $G(\Bbbk_n/\mathbb{Q})$; compare Corollary 1.14. Since λ is a PV-number, the last observation shows that

$$\xi := \max\left\{|\sigma_j(\lambda)| \,\middle|\, j \in \{2, \dots, \phi(n)/2\}\right\} \in (0,1)\,.$$

The assertion follows. □

The following result on a simple relation between cyclotomic model sets of the form $\varLambda_n(t,W) \in \mathcal{M}(\mathcal{O}_n)$, where $n \in \mathbb{N} \setminus \{1,2\}$, and the $\mathbb{Q}$-span $\mathbb{Q}\mathcal{O}_n = \mathbb{K}_n$ of their underlying dense $\mathbb{Z}$-module $\mathcal{O}_n$, will play a key role. Loosely, it says that $\varLambda_n(t,W)$ 'sees' every finite subset of $\mathbb{K}_n$, but possibly on a different scale.

LEMMA 1.81. *Let $n \in \mathbb{N} \setminus \{1,2\}$ and let $\varLambda_n(t,W) \in \mathcal{M}(\mathcal{O}_n)$ be a cyclotomic model set. Then, for any finite set $F \subset \mathbb{K}_n$, there is a homothety $h \colon \mathbb{R}^2 \longrightarrow \mathbb{R}^2$ such that $h[F] \subset \varLambda_n(t,W)$. In particular, if $t = 0$ and $0 \in \operatorname{int}(W)$, there is even a dilatation $d \colon \mathbb{R}^2 \longrightarrow \mathbb{R}^2$ such that $d[F] \subset \varLambda_n(0,W)$.*

PROOF. Without loss of generality, we may assume that F is non-empty. We consider the $\mathbb{Q}$-coordinates of the elements of F with respect to the $\mathbb{Q}$-basis $\{1, \zeta_n, \zeta_n^2, \dots, \zeta_n^{\phi(n)-1}\}$ of $\mathbb{K}_n$ (cf. Proposition 1.11) and let $l \in \mathbb{N}$ be the least common multiple of all their denominators. Then, by Remark 1.24, we get $lF \subset \mathcal{O}_n$. If $n \in \{3,4,6\}$, we are done by defining the homothety $h \colon \mathbb{R}^2 \longrightarrow \mathbb{R}^2$ by

$$z \longmapsto lz + t\,.$$

Secondly, suppose that $n \notin \{3,4,6\}$. Let $.^{\star_n}$ be the star map that is used in the construction of $\varLambda_n(t,W)$. From $\operatorname{int}(W) \neq \varnothing$ and the denseness of $[\mathcal{O}_n]^{\star_n}$ in $(\mathbb{R}^2)^{\phi(n)/2-1}$, there follows the existence of a suitable $z_0 \in \mathcal{O}_n$ with $z_0{}^{\star_n} \in \operatorname{int}(W)$. Consider the open neighbourhood

$$V := \operatorname{int}(W) - z_0{}^{\star_n}$$

of 0 in $(\mathbb{R}^2)^{\phi(n)/2-1}$. Next, we choose a PV-number λ of degree $\phi(n)/2$ in $\mathcal{o}_n$; compare Lemma 1.29. Consider the $\mathbb{Z}$-module endomorphism

$$(m_\lambda^{(n)})^{\star_n} \colon [\mathcal{O}_n]^{\star_n} \longrightarrow [\mathcal{O}_n]^{\star_n}\,,$$

as defined in Definition 1.77(c). Since λ is a PV-number, Lemma 1.80 shows that $(m_\lambda^{(n)})^{\star_n}$ is contractive (in the sense which was made precise in that lemma). Since all norms on $(\mathbb{R}^2)^{\phi(n)/2-1}$ are equivalent, it follows the existence of a suitable $k \in \mathbb{N}$ such that

$$\big((m_\lambda^{(n)})^{\star_n}\big)^k\big[\,[lF]^{\star_n}\big] \subset V\,.$$

It follows that $\big\{(\lambda^k z + z_0)^{\star_n} \,|\, z \in lF\big\} \subset \operatorname{int}(W)$ and further that $h[F] \subset \varLambda_n(t,W)$, where $h\colon \mathbb{R}^2 \longrightarrow \mathbb{R}^2$ is the homothety given by

$$z \longmapsto (l\lambda^k)z + (z_0 + t)\,.$$

The additional statement follows immediately from the foregoing proof in connection with the trivial observation that $0 \in \mathcal{O}_n$ maps, under the star map $.^{\star_n}$, to $0 \in (\mathbb{R}^2)^{\phi(n)/2-1}$. □

REMARK 1.82. As the general structure shows, the restriction in Lemma 1.81 to cyclotomic model sets is by no means necessary. In fact, the related general algebraic setting of **[51]** can be used to extend this result also to higher dimensions. There, the role of the PV-numbers (or PV-units) will be taken by suitable hyperbolic lattice endomorphisms (or automorphisms); see Lemma 1.108.

LEMMA 1.83. *Cyclotomic model sets are algebraic Delone sets.*

PROOF. Let $n \in \mathbb{N}\setminus\{1,2\}$ and let $\varLambda_n(t,W) \in \mathcal{M}(\mathcal{O}_n)$ be a cyclotomic model set. First, by Remark 1.71, $\varLambda_n(t,W)$ is a Delone set. Secondly, since $\varLambda_n(t,W) - \varLambda_n(t,W) \subset \mathcal{O}_n$ and $\overline{\mathcal{O}_n} = \mathcal{O}_n$, one has

$$\mathbb{K}_{\varLambda_n(t,W)} \;\subset\; \mathbb{Q}(\mathcal{O}_n) \;\subset\; \mathbb{Q}(\mathbb{K}_n) \;=\; \mathbb{K}_n\,. \tag{1.21}$$

Consequently, since $[\mathbb{K}_n : \mathbb{Q}] = \phi(n)$ by Proposition 1.11, one has

$$\big[\mathbb{K}_{\varLambda_n(t,W)} : \mathbb{Q}\big] \;\leq\; [\mathbb{K}_n : \mathbb{Q}] \;=\; \phi(n) \;<\; \infty\,.$$

It follows that $\varLambda_n(t,W)$ satisfies property (A) of Definition 1.61, whereas property (B) follows immediately from Lemma 1.81 in conjunction with Relation (1.21). □

As an easy application of Lemma 1.81, one obtains the following result on the set of $\varLambda_n(t,W)$-directions (resp., 1-dimensional $\varLambda_n(t,W)$-subspaces).

LEMMA 1.84. *Let $n \in \mathbb{N}\setminus\{1,2\}$ and let $\varLambda_n(t,W) \in \mathcal{M}(\mathcal{O}_n)$ be a cyclotomic model set. Then, one has:*

(a) *The set of $\mathcal{O}_n$-directions is precisely the set of $\varLambda_n(t,W)$-directions.*
(b) *The set of 1-dimensional $\mathcal{O}_n$-subspaces is precisely the set of 1-dimensional $\varLambda_n(t,W)$-subspaces.*

PROOF. Let us start with (a). Since one has $\varLambda_n(t,W) - \varLambda_n(t,W) \subset \mathcal{O}_n$, every $\varLambda_n(t,W)$-direction is an $\mathcal{O}_n$-direction. For the converse, let $u \in \mathbb{S}^1$ be an $\mathcal{O}_n$-direction, say parallel to $o \in \mathcal{O}_n \setminus \{0\}$. By Lemma 1.81, there is a homothety $h\colon \mathbb{R}^2 \longrightarrow \mathbb{R}^2$ such that $h[\{0,o\}] \subset \varLambda_n(t,W)$. It follows that $h(o) - h(0) \in (\varLambda_n(t,W) - \varLambda_n(t,W)) \setminus \{0\}$. Since $h(o) - h(0)$ is parallel to o, the assertion follows. Part (b) follows from similar arguments. □

REMARK 1.85. Lemma 1.84 shows that the notion of $\mathcal{O}_n$-directions in the context of cyclotomic model sets is a natural extension of the notion of *lattice directions* (i.e., $\mathbb{Z}^d$-directions, $d \geq 2$) in **[33]**.

REMARK 1.86. It is well known that the nth cyclotomic field $\mathbb{K}_n$ (excluding $\mathbb{Q}$) has class number one if and only if

$$n \in \{3,4,5,7,8,9,11,12,13,15,16,17,19,20,21,\\ 24,25,27,28,32,33,35,36,40,44,45,48,60,84\}, \tag{1.22}$$

where $n \not\equiv 2 \pmod 4$ to avoid double counting and $n = 1$ (corresponding to $\mathbb{K}_1 = \mathbb{Q}$) is excluded, because we shall not need it for our applications; see [**66**, Theorem 11.1]. It follows that, with the restriction $n \not\equiv 2 \pmod 4$, the values of n in (1.22) correspond to all cases (except $n = 1$), where $\mathcal{O}_n$ is a principal ideal domain (PID) and thus is a unique factorization domain (UFD). Moreover, it is known that, in this cyclotomic situation, the unique prime factorization property of $\mathcal{O}_n$ implies that of o_n; see [**66**, Proposition 11.19].

DEFINITION 1.87. For $S \subset \mathbb{C}$, we set $\sqrt{S} := \{x \in \mathbb{C} \mid x^2 \in S\}$.

REMARK 1.88. Note that, for $z \in \mathbb{C}$, say $z = re^{i\phi}$, where $r = \|z\|$ and $\phi \in [0, 2\pi)$, one has $\sqrt{\{z\}} = \{\sqrt{r}e^{i\phi/2}, -\sqrt{r}e^{i\phi/2}\}$.

Since it can be easily seen that, for $n \in \mathbb{N}$, the set of $\mathcal{O}_n$-directions actually is a *subgroup* of the group $(\mathbb{S}^1, \cdot)$ of all directions in the plane, one might be interested in the structure of the group of $\mathcal{O}_n$-directions. The next result sheds some light on this issue, where we exclude the trivial cases $n \in \{1,2\}$.

PROPOSITION 1.89. *The following statements hold:*

(a) *For $n \in \mathbb{N} \setminus \{1,2\}$, one has*

$$\{u^2 \mid u \in \mathbb{S}^1 \textit{ is an } \mathcal{O}_n\textit{-direction}\} = \mathbb{K}_n \cap \mathbb{S}^1,$$

whence

$$\{u \in \mathbb{S}^1 \mid u \textit{ is an } \mathcal{O}_n\textit{-direction}\} = \sqrt{\mathbb{K}_n \cap \mathbb{S}^1}.$$

(b) *If n is one of the values from the list (1.22), one has the isomorphies of groups*

$$\{u^2 \mid u \in \mathbb{S}^1 \textit{ is an } \mathcal{O}_n\textit{-direction}\} \cong \mathsf{C}_{N(n)} \times \mathbb{Z}^{(\aleph_0)}$$

and

$$\{u \in \mathbb{S}^1 \mid u \textit{ is an } \mathcal{O}_n\textit{-direction}\} \cong \mathsf{C}_{2N(n)} \times \mathbb{Z}^{(\aleph_0)},$$

where $N(n)$ is the function from (1.2), $\mathbb{Z}^{(\aleph_0)}$ stands for the direct sum of countably infinitely many infinite cyclic groups and, for $m \in \mathbb{N}$, C_m denotes the cyclic group of order m.

PROOF. For part (a), let $u \in \mathbb{S}^1$ be an $\mathcal{O}_n$-direction, say $u = \alpha/\|\alpha\| = \alpha/\sqrt{\alpha\bar{\alpha}}$, where $\alpha \in \mathcal{O}_n \setminus \{0\}$. It follows that $u^2 = \alpha^2/(\alpha\bar{\alpha}) = \alpha/\bar{\alpha} \in \mathbb{K}_n \cap \mathbb{S}^1$. For the reverse inclusion, let $z \in \mathbb{K}_n \cap \mathbb{S}^1$. Since $((\zeta_n - \bar{\zeta}_n)/\|\zeta_n - \bar{\zeta}_n\|)^2 = -1$, one may assume, without loss of generality, that $z \neq -1$. Next, consider a square root

$$\sqrt{z} := \frac{1+z}{\|1+z\|} = \frac{1+z}{\sqrt{(1+z)\overline{(1+z)}}} \in \mathbb{S}^1$$

of z. It follows that

$$z = (\sqrt{z})^2 = \frac{\sqrt{z}}{\overline{\sqrt{z}}} = \frac{\frac{1+z}{\sqrt{(1+z)\overline{(1+z)}}}}{\frac{\overline{1+z}}{\sqrt{(1+z)\overline{(1+z)}}}} = \frac{1+z}{\overline{1+z}},$$

and, since $(1+z)/(\overline{1+z}) = \alpha/\bar{\alpha}$ for a suitable $\alpha \in \mathcal{O}_n \setminus \{0\}$, one obtains

$$z = \frac{\alpha}{\bar{\alpha}} = \left(\frac{\alpha}{\sqrt{\alpha\bar{\alpha}}}\right)^2 = \left(\frac{\alpha}{\|\alpha\|}\right)^2 .$$

Claim (a) follows. For part (b), the first isomorphy on the set of squares of $\mathcal{O}_n$-directions follows immediately from part (a) in conjunction with the results in [**53**]. In fact, it was shown there that, for all n from the list (1.22), one has the isomorphy

$$\mathbb{K}_n \cap \mathbb{S}^1 \cong \mathsf{C}_{N(n)} \times \mathbb{Z}^{(\aleph_0)} .$$

More precisely, it was shown in [**53**] that every element $z \in \mathbb{K}_n \cap \mathbb{S}^1$ is of the form $z = v/w$, where $v, w \in \mathcal{O}_n \setminus \{0\}$ are coprime with

$$N_{\mathbb{K}_n/\Bbbk_n}(v) = N_{\mathbb{K}_n/\Bbbk_n}(w) =: \nu \in \mathcal{o}_n .$$

Moreover, every $z \in \mathbb{K}_n \cap \mathbb{S}^1$ can be uniquely written as

$$z = \varepsilon \prod_{\mathfrak{p}} \left(\frac{\omega_{\mathfrak{p}}}{\bar{\omega}_{\mathfrak{p}}}\right)^{t_{\mathfrak{p}}} ,$$

where $\mathfrak{p}$ runs over the primes of $\mathcal{o}_n$ that divide ν from above and split into $\mathcal{O}_n$-primes as $\mathfrak{p} = \omega_{\mathfrak{p}}\bar{\omega}_{\mathfrak{p}}$ with $\omega_{\mathfrak{p}}/\bar{\omega}_{\mathfrak{p}} \notin \mathcal{O}_n^{\times}$ in the extension from $\Bbbk_n$ to $\mathbb{K}_n$, ε is an $N(n)$th root of unity and the $t_{\mathfrak{p}}$'s are elements of $\mathbb{Z}$. By taking square roots, it follows that every $\mathcal{O}_n$-direction $u \in \mathbb{S}^1$ can be uniquely written as

$$z = \varepsilon' \prod_{\mathfrak{p}} \left(\frac{\omega_{\mathfrak{p}}}{\sqrt{\mathfrak{p}}}\right)^{t_{\mathfrak{p}}} ,$$

where the $\mathfrak{p}$'s and the $t_{\mathfrak{p}}$'s are as above and ε' is a $(2\,N(n))$th root of unity. Taking the $(\omega_{\mathfrak{p}}/\sqrt{\mathfrak{p}})$'s as the new generators of the groups of type $\mathbb{Z}$, the second isomorphy of claim (b) follows. This completes the proof. □

We shall also need the following consequence of Weyl's theory of uniform distribution, which is concerned with an analytical aspect of regular cyclotomic model sets. This was analyzed in general in [**60, 49**]. We need the following variant.

THEOREM 1.90. *Let $n \in \mathbb{N} \setminus \{1,2,3,4,6\}$ and let $\varLambda := \varLambda_n(t, W) \in \mathcal{M}(\mathcal{O}_n)$ be a regular (aperiodic) cyclotomic model set, say constructed by use of the star map $.^{\star_n}$. Then, for all $a \in \mathbb{R}^2$, one has the identity*

$$\lim_{R\to\infty} \frac{1}{\operatorname{card}(\varLambda \cap B_R(a))} \sum_{x \in \varLambda \cap B_R(a)} x^{\star_n} = \frac{1}{\operatorname{vol}(W)} \int_W y \,\mathrm{d}\lambda(y) ,$$

where λ denotes the Lebesgue measure on $(\mathbb{R}^2)^{\frac{\phi(n)}{2}-1}$.

PROOF. The assertion follows from the multi-dimensional version of [**48**, Theorem 3]. □

REMARK 1.91. As it is the case for cyclotomic model sets of the form $\varLambda_5(t, W) \in \mathcal{M}(\mathcal{O}_5)$, also the so-called *Penrose model sets* can be described as model sets with underlying $\mathbb{Z}$-module $\mathcal{O}_5$. More precisely, let C_5 denote the cyclic group of order 5, i.e., $\mathsf{C}_5 = \mathbb{Z}/5\mathbb{Z}$. Moreover, C_5

is understood to be supplied with the discrete topology. By Remark 1.24, every $z \in \mathcal{O}_5$ can be written as

$$z = \sum_{j=0}^{3} a_j(z)\, \zeta_5^j\,,$$

for uniquely determined $a_j(z) \in \mathbb{Z}$, $j \in \{0, \dots, 3\}$. Let $\sigma_2 \in G(\mathbb{K}_5/\mathbb{Q})$ be given by $\zeta_5 \longmapsto \zeta_5^2$. The latter gives rise to a map $.^{\sim} : \mathcal{O}_5 \longrightarrow \mathbb{R}^2 \times (\mathbb{R}^2 \times \mathsf{C}_5)$, defined by

$$z \longmapsto \Big(z, \big(\sigma_2(z), \sum_{j=0}^{3}(a_j(z) \;(\operatorname{mod} 5))\big)\Big)\,.$$

Via projection on the second factor, this induces a map $.^{\star} \colon \mathcal{O}_5 \longrightarrow \mathbb{R}^2 \times \mathsf{C}_5$. Then, $[\mathcal{O}_5]^{\sim}$ is a lattice in $\mathbb{R}^2 \times (\mathbb{R}^2 \times \mathsf{C}_5)$ and $[\mathcal{O}_5]^{\star}$ is dense in $\mathbb{R}^2 \times \mathsf{C}_5$. The class of *Penrose model sets* arises from the following cut and project scheme.

$$\begin{array}{ccccc} \mathbb{R}^2 & \stackrel{\pi}{\longleftarrow} & \mathbb{R}^2 \times (\mathbb{R}^2 \times \mathsf{C}_5) & \stackrel{\pi_{\text{int}}}{\longrightarrow} & \mathbb{R}^2 \times \mathsf{C}_5 \\ \cup \ \text{dense} & & \cup \ \text{lattice} & & \cup \ \text{dense} \\ \mathcal{O}_5 & \stackrel{1-1}{\longleftrightarrow} & [\mathcal{O}_5]^{\sim} & \stackrel{1-1}{\longleftrightarrow} & [\mathcal{O}_5]^{\star} \end{array} \tag{1.23}$$

Note that, although the proofs are slightly more technically, many of the results presented in this thesis also hold in the case of Penrose model sets $\varLambda$. Note also that the class of Penrose model sets contains the vertex sets of the famous fivefold symmetric rhombic Penrose tilings; see [**9**] for a summary.

1.2.3.3. *Icosahedral model sets.* Let $\mathbb{H}$ be the skew field of *Hamiltonian quaternions*, i.e.,

$$\mathbb{H} = \{a + bi + cj + dk \,|\, a, b, c, d \in \mathbb{R}\}\,,$$

a 4-dimensional vector space over $\mathbb{R}$ with a non-commutative multiplication determined by

$$i^2 = j^2 = k^2 = ijk = -1\,.$$

The *conjugate* of $\alpha = a + bi + cj + dk \in \mathbb{H}$ is defined by $\bar{\alpha} = a - bi - cj - dk$, the *reduced norm* by $\operatorname{nr}(\alpha) = \alpha\bar{\alpha} = a^2 + b^2 + c^2 + d^2$ and the *reduced trace* by $\operatorname{tr}(\alpha) = \alpha + \bar{\alpha} = 2a$. Moreover, we shall sometimes call $\operatorname{Re}(\alpha) := a \in \mathbb{R}$ the *real part* and $\operatorname{Im}(\alpha) := (b, c, d)^t \in \mathbb{R}^3$ the *imaginary part* of α. Let $\mathbb{H}_0$ be the set of quaternions with real part 0, i.e.,

$$\mathbb{H}_0 := \{\alpha \in \mathbb{H} \,|\, \operatorname{tr}(\alpha) = 0\} = \{bi + cj + dk \,|\, b, c, d \in \mathbb{R}\} \;\cong\; \mathbb{R}^3\,.$$

The *icosian ring* $\mathbb{I}$ (whose members are called *icosians*) is the additive subgroup of $\mathbb{H}$ that is generated by the quaternions

$$\big((\pm 1, 0, 0, 0)^t\big)^{\mathsf{A}}, \tfrac{1}{2}\,\big((\pm 1, \pm 1, \pm 1, \pm 1)^t\big)^{\mathsf{A}}, \tfrac{1}{2}\,\big((0, \pm 1, \pm\tau', \tau)^t\big)^{\mathsf{A}}\,,$$

where we identify $\mathbb{H}$ with $\mathbb{R}^4$ via the basis $\{1, i, j, k\}$ and, as in [**23**, Chapter 8], the superscript A indicates that all even permutations of the coordinates are allowed. Moreover, τ' arises from τ via the unique non-trivial Galois automorphism $.'$ on $\Bbbk_5 = \mathbb{Q}(\tau)$ (given by $\sqrt{5} \longmapsto -\sqrt{5}$), i.e., $\tau' = -1/\tau = 1 - \tau$. Note that $\mathbb{I}$ is a ring, because these generators (which have reduced norm 1) form a multiplicative group, the *icosian group*, of order 120. Note further that $\mathcal{O}_5 = \mathbb{Z}[\tau]$ and that $\mathbb{I}$ is also a free $\mathcal{O}_5$-module of rank 4. In fact, $\mathbb{I}$ is a maximal order of the quaternion algebra $\mathbb{H}(\mathbb{k}_5)$ over $\mathbb{k}_5$, defined as

$$\mathbb{H}(\mathbb{k}_5) = \{a + bi + cj + dk \mid a, b, c, d \in \mathbb{k}_5\};$$

see [15]. The set

$$\mathbb{I}_0 := \mathrm{Im}[\mathbb{I} \cap \mathbb{H}_0] \subset \mathbb{R}^3$$

of 'pure imaginary' icosians is generated as an additive group by the elements

$$\left((\pm 1, 0, 0)^t\right)^{\mathsf{A}}, \tfrac{1}{2}\left((\pm 1, \pm\tau', \pm\tau)^t\right)^{\mathsf{A}},$$

where the superscript A is defined as above. Consider the standard body centred icosahedral module $\mathcal{M}_{\mathrm{B}}$ of quasicrystallography, defined as

$$\begin{aligned}
\mathcal{M}_{\mathrm{B}} &:= \left\langle\{(2,0,0)^t, (1,1,1)^t, (\tau,0,1)^t\}\right\rangle_{\mathcal{O}_5} \\
&= \left\langle\{(0,2,0)^t, (-1,-\tau',\tau)^t, (1,1,1)^t\}\right\rangle_{\mathcal{O}_5} \qquad (1.24)\\
&= \left\{(\beta,\gamma,\delta)^t \;\middle|\; \begin{array}{l}\beta,\gamma,\delta \in \mathcal{O}_5, \text{ with} \\ \tau^2\beta + \tau\gamma + \delta \equiv 0 \pmod 2\end{array}\right\}.
\end{aligned}$$

One has $\mathrm{Im}[\mathbb{I}] = \frac{1}{2}\mathcal{M}_{\mathrm{B}}$ and, further, $\mathbb{I}_0 = \frac{1}{2}\mathcal{M}_{\mathrm{F}}$, where $\mathcal{M}_{\mathrm{F}}$ is the standard face centred icosahedral module of quasicrystallography, defined as

$$\begin{aligned}
\mathcal{M}_{\mathrm{F}} &:= \left\{(\beta,\gamma,\delta)^t \;\middle|\; \begin{array}{l}\beta,\gamma,\delta \in \mathcal{O}_5, \text{ with} \\ \beta \equiv \tau\gamma \equiv \tau^2\delta \pmod 2\end{array}\right\}, \\
&= \left\{(\beta,\gamma,\delta)^t \in \mathcal{M}_{\mathrm{B}} \mid \beta + \gamma + \delta \equiv 0 \pmod 2\right\}, \qquad (1.25)\\
&= \left\langle\{(2,0,0)^t, (\tau+1,\tau,1)^t, (0,0,2)^t\}\right\rangle_{\mathcal{O}_5} \overset{4}{\subset} \mathcal{M}_{\mathrm{B}} \subset \mathbb{R}^3,
\end{aligned}$$

where $\mathcal{M}_{\mathrm{F}} \overset{4}{\subset} \mathcal{M}_{\mathrm{B}}$ denotes the fact that $\mathcal{M}_{\mathrm{F}}$ is a subgroup of index 4 in $\mathcal{M}_{\mathrm{B}}$; see [4] and references therein. Both $\mathcal{M}_{\mathrm{B}}$ and $\mathcal{M}_{\mathrm{F}}$ are free $\mathcal{O}_5$-modules of rank 3, and are hence $\mathbb{Z}$-modules of rank 6. Moreover, both $\mathcal{M}_{\mathrm{B}}$ and $\mathcal{M}_{\mathrm{F}}$ have icosahedral symmetry, i.e., they are invariant under the action of the rotation group Y. This group is generated by the rotations which are given, with respect to the canonical basis, by the following matrices

$$\begin{pmatrix} -1 & 0 & 0 \\ 0 & -1 & 0 \\ 0 & 0 & 1 \end{pmatrix}, \quad \frac{1}{2}\begin{pmatrix} \tau & -1 & -\tau' \\ 1 & -\tau' & -\tau \\ -\tau' & \tau & 1 \end{pmatrix}. \qquad (1.26)$$

Note that Y is the rotation group of the regular icosahedron centred at the origin $0 \in \mathbb{R}^3$ with orientation such that each coordinate axis passes through the mid-point of an edge, thus coinciding with 2-fold axes of the icosahedron. Moreover, the matrix on the left (resp., on the right) is an order 2 (resp., order 5) rotation. We let

$$\begin{aligned}
.^\star\colon \mathrm{Im}[\mathbb{I}] &\longrightarrow \mathbb{R}^3, \\
\alpha &\longmapsto \alpha^\star,
\end{aligned}$$

be the map defined by applying the conjugation $.'$ to each coordinate of α. Furthermore, let the map

$$.^{\sim} : \mathrm{Im}[\mathbb{I}] \longrightarrow \mathbb{R}^3 \times \mathbb{R}^3$$

be given by

$$\alpha \longmapsto (\alpha, \alpha^\star)\,.$$

Note that the image $[\mathrm{Im}[\mathbb{I}]]^{\sim}$ is a lattice in $\mathbb{R}^3 \times \mathbb{R}^3 \cong \mathbb{R}^6$ and, moreover, that it has a natural interpretation as a weight lattice of type D_6^*; cf. **[21, 23]** for background. Further, note that both $\mathrm{Im}[\mathbb{I}]$ and the image $[\mathrm{Im}[\mathbb{I}]]^\star$ are dense in $\mathbb{R}^3$.

This provides the following cut and project scheme.

$$\begin{array}{ccccc} \mathbb{R}^3 & \stackrel{\pi}{\longleftarrow} & \mathbb{R}^3 \times \mathbb{R}^3 & \stackrel{\pi_{\mathrm{int}}}{\longrightarrow} & \mathbb{R}^3 \\ \cup_{\text{dense}} & & \cup_{\text{lattice}} & & \cup_{\text{dense}} \\ \mathrm{Im}[\mathbb{I}] & \stackrel{1-1}{\longleftrightarrow} & [\mathrm{Im}[\mathbb{I}]]^{\sim} & \stackrel{1-1}{\longleftrightarrow} & [\mathrm{Im}[\mathbb{I}]]^\star \end{array} \tag{1.27}$$

REMARK 1.92. Similarly, $\mathbb{I}_0$ gives rise to a cut and project scheme, where the lattice $[\mathbb{I}_0]^{\sim}$ in $\mathbb{R}^3 \times \mathbb{R}^3$ has a natural interpretation as a root lattice of type D_6; cf. **[21, 23]** again. Note also that there is another $\mathbb{Z}$-module of rank 6, intermediate between $\mathcal{M}_{\mathrm{F}}$ and $\mathcal{M}_{\mathrm{B}}$, which also has icosahedral symmetry. This is $\mathcal{M}_{\mathrm{P}}$, defined as

$$\mathcal{M}_{\mathrm{P}} := \left\{(\beta,\gamma,\delta)^t \in \mathcal{M}_{\mathrm{B}} \mid \beta+\gamma+\delta \equiv 0 \text{ or } \tau \pmod 2\right\}\,.$$

In contrast to $\mathcal{M}_{\mathrm{F}}$ and $\mathcal{M}_{\mathrm{B}}$, $\mathcal{M}_{\mathrm{P}}$ fails to be an $\mathcal{O}_5$-module. In fact, $\mathcal{M}_{\mathrm{P}}$ is a $\mathbb{Z}[2\tau]$-module only.

DEFINITION 1.93. Given any window $W \subset \mathbb{R}^3$ and any $t \in \mathbb{R}^3$, we obtain a three-dimensional model set, a so-called *icosahedral model set*,

$$\varLambda_{\mathrm{ico}}(t,W) := t + \varLambda_{\mathrm{ico}}(W)$$

relative to the above cut and project scheme (1.27) by setting

$$\varLambda_{\mathrm{ico}}(W) := \{\alpha \in \mathrm{Im}[\mathbb{I}] \mid \alpha^\star \in W\}\,.$$

Further, we denote by $\mathcal{I}$ the set of all icosahedral model sets and define $\mathcal{I}_g$ as the subset of $\mathcal{I}$ consisting of all generic icosahedral model sets. Further, for a window $W \subset \mathbb{R}^3$, the elements of the subset

$$\{\varLambda_{\mathrm{ico}}(t,\tau+W) \mid t,\tau \in \mathbb{R}^3\} \,\cap\, \mathcal{I}_g$$

of $\mathcal{I}_g$ are called *$\mathcal{I}_g(W)$-sets.*

REMARK 1.94. Note that icosahedral model sets $\varLambda := \varLambda_{\mathrm{ico}}(t,W) \subset \mathbb{R}^3$ are aperiodic. If $\varLambda$ is both generic and regular, and, if a suitable translate of the window W has full icosahedral symmetry (i.e., if a suitable translate of the window W is invariant under the action of the group $Y_h^\star$ of order 120, where $Y_h^\star := Y^\star \dot{\cup} (-Y^\star)$ and $Y^\star$ is the group of rotations of order 60 generated by the two matrices that arise from the two matrices in (1.26) by applying the conjugation $.'$ to each entry), then $\varLambda$ has full icosahedral symmetry $Y_h := Y \dot{\cup} (-Y)$ in the sense of symmetries of LI-classes. Typical examples are balls and suitably oriented versions of the icosahedron, the dodecahedron, the rhombic triacontahedron (the latter also known as Kepler's body) and its dual, the icosidodecahedron.

We shall now prove that icosahedral model sets $\varLambda$ can be nicely 'sliced' into cyclotomic model sets, where the slices are translates of the hyperplane in $\mathbb{R}^3$ orthogonal to the vector $(\tau,0,1)^t$. The latter observation will be crucial, since it brings us later into the position to use our results on discrete tomography of cyclotomic model sets, slice by slice.

DEFINITION 1.95. Let $(a,b,c)^t \in \mathbb{R}^3 \setminus \{0\}$. We use $H^{(a,b,c)}$ to denote the hyperplane in $\mathbb{R}^3$ that is the orthogonal complement of the line $\mathbb{R}(a,b,c)^t$, i.e.,

$$H^{(a,b,c)} := \left(\mathbb{R}(a,b,c)^t\right)^{\perp}.$$

LEMMA 1.96. *The following equations hold:*

(a) $\mathrm{Im}[\mathbb{I}] \cap H^{(\tau,0,1)} = \left\langle (0,1,0)^t, \frac{1}{2}(-1,-\tau',\tau)^t \right\rangle_{\mathcal{O}_5}$.

(b) $\left[\mathrm{Im}[\mathbb{I}] \cap H^{(\tau,0,1)}\right]^\star = [\mathrm{Im}[\mathbb{I}]]^\star \cap H^{(\tau',0,1)}$.

PROOF. Part (a) follows from Equation (1.24) and the relation $\mathrm{Im}[\mathbb{I}] = \frac{1}{2}\mathcal{M}_{\mathrm{B}}$. Part (b) follows from part (a) in conjunction with the identity $((\tau,0,1)^t)^\star = (\tau',0,1)^t$. □

DEFINITION 1.97.

(a) We denote by Φ the $\mathbb{R}$-linear map $\Phi\colon H^{(\tau,0,1)} \longrightarrow \mathbb{C}$, defined by

$$r(0,1,0)^t + s\tfrac{1}{2}(-1,-\tau',\tau)^t \longmapsto r + s\,\zeta_5\,.$$

(b) We denote by $\Phi^\star$ the $\mathbb{R}$-linear map $\Phi^\star\colon H^{(\tau',0,1)} \longrightarrow \mathbb{C}$, defined by

$$r(0,1,0)^t + s\tfrac{1}{2}(-1,-\tau,\tau')^t \longmapsto r + s\,\zeta_5^3\,.$$

REMARK 1.98. Note that $H^{(\tau,0,1)}$, $H^{(\tau',0,1)}$ and $\mathbb{C}$ are two-dimensional Euclidean vector spaces, the first and the second one with respect to the inner product induced by the canonical inner product on $\mathbb{R}^3$, and the third one with respect to the canonical inner product on $\mathbb{R}^2 \cong \mathbb{C}$. Further, Φ and $\Phi^\star$ preserve these inner products, the latter being equivalent to the fact that Φ and $\Phi^\star$ are isometries of Euclidean vector spaces. More concretely, one has the identities

$$\|r(0,1,0)^t + s\tfrac{1}{2}(-1,-\tau',\tau)^t\| = |r + s\,\zeta_5| = \sqrt{r^2+s^2-rs\tau'}$$

and

$$\|r(0,1,0)^t + s\tfrac{1}{2}(-1,-\tau,\tau')^t\| = |r + s\,\zeta_5^3| = \sqrt{r^2+s^2-rs\tau}\,.$$

In particular, Φ and $\Phi^\star$ are bijections. Moreover, if one identifies $\mathbb{C}$ with the xy-plane in $\mathbb{R}^3$ in the canonical way, Φ (resp., $\Phi^\star$) has a unique extension to a direct rigid motion of $\mathbb{R}^3$, i.e., an element of the special orthogonal group $SO(3,\mathbb{R})$. Recall from Lemma 1.9 that the ring $\mathcal{O}_5 = \mathbb{Z}[\zeta_5]$ is an $\mathcal{o}_5$-module of rank two; more precisely, it was shown there that $\mathcal{O}_5 = \mathcal{o}_5 + \mathcal{o}_5\zeta_5$. Since ζ_5^3 is also a primitive 5th root of unity, one also has the equality $\mathcal{O}_5 = \mathbb{Z}[\zeta_5^3] = \mathcal{o}_5 + \mathcal{o}_5\zeta_5^3$; see Lemma 1.9 again.

LEMMA 1.99. *Via restriction, the map Φ induces an isomorphism of rank two $\mathcal{o}_5$-modules*

$$\mathrm{Im}[\mathbb{I}] \cap H^{(\tau,0,1)} \xrightarrow{\;\Phi\;} \mathcal{O}_5\,.$$

Likewise, the map $\Phi^\star$ induces an isomorphism of rank two $\mathcal{o}_5$-modules

$$[\mathrm{Im}[\mathbb{I}]]^\star \cap H^{(\tau',0,1)} \xrightarrow{\;\Phi^\star\;} \mathcal{O}_5\,.$$

PROOF. This follows immediately from Lemma 1.96 and Remark 1.98. □

COROLLARY 1.100. *The set*

$$B_{\mathrm{Im}[\mathbb{I}]} := \{\tfrac{1}{2}(1,1,1)^t, \tfrac{\tau}{2}(1,1,1)^t, (0,1,0)^t, \tfrac{1}{2}(-1,-\tau',\tau)^t, \tfrac{1}{2}(\tau',-\tau,1)^t, \tfrac{1}{2}(-\tau',-\tau,-1)^t\}.$$

is simultaneously a $\mathbb{Z}$*-basis of* $\mathrm{Im}[\mathbb{I}]$ *and a* $\mathbb{Q}$*-basis of* $\langle \mathrm{Im}[\mathbb{I}]\rangle_{\mathbb{Q}} = \mathbb{Q}\,\mathrm{Im}[\mathbb{I}] = (\mathbb{k}_5)^3$.

PROOF. First, observe that one has the equalities

$$\Phi((0,1,0)^t) = 1\,,\ \Phi(\tfrac{1}{2}(-1,-\tau',\tau)^t) = \zeta_5\,,\ \Phi(\tfrac{1}{2}(\tau',-\tau,1)^t) = \zeta_5^2$$

and $\Phi(\frac{1}{2}(-\tau',-\tau,-1)^t) = \zeta_5^3$. Then, the fact that $B_{\mathrm{Im}[\mathbb{I}]}$ is simultaneously a $\mathbb{Z}$-basis of $\mathrm{Im}[\mathbb{I}]$ and a $\mathbb{Q}$-basis of $\langle \mathrm{Im}[\mathbb{I}]\rangle_{\mathbb{Q}} = \mathbb{Q}\,\mathrm{Im}[\mathbb{I}]$ follows immediately from Lemma 1.99 and Equation (1.24) in conjunction with the fact that the set $\{1,\zeta_5,\zeta_5^2,\zeta_5^3\}$ is simultaneously a $\mathbb{Z}$-basis of the $\mathbb{Z}$-module $\mathcal{O}_5$ and a $\mathbb{Q}$-basis of $\mathbb{K}_5 = \langle \mathcal{O}_5\rangle_{\mathbb{Q}} = \mathbb{Q}\mathcal{O}_5$; cf. Proposition 1.11 and Remark 1.24. The equality $(\mathbb{k}_5)^3 = \langle \mathrm{Im}[\mathbb{I}]\rangle_{\mathbb{Q}}$ then follows immediately from the obvious inclusion $\langle \mathrm{Im}[\mathbb{I}]\rangle_{\mathbb{Q}} \subset (\mathbb{k}_5)^3$ in conjunction with the fact that both sets are 6-dimensional vector spaces over $\mathbb{Q}$; see Corollary 1.14. □

PROPOSITION 1.101. *Let* $\varLambda_{\mathrm{ico}}(t,W)$ *be an icosahedral model set. Then, for every* $\alpha \in \mathrm{Im}[\mathbb{I}]$*, one has the identity*

$$\Phi\left[\left(\varLambda_{\mathrm{ico}}(t,W) \,\cap\, \left(t+\alpha+H^{(\tau,0,1)}\right)\right) - t - \alpha\right] = \left\{z \in \mathcal{O}_5 \,\middle|\, z^{\star_5} \in W_\alpha\right\},$$

where $.^{\star_5}$ *is the Galois automorphism in* $G(\mathbb{K}_5/\mathbb{Q})$*, defined by* $\zeta_5 \longmapsto \zeta_5^3$ *and*

$$W_\alpha := \Phi^\star\left[\left(W \,\cap\, \left(\alpha^\star + H^{(\tau',0,1)}\right)\right) - \alpha^\star\right]. \tag{1.28}$$

PROOF. First, consider $\Phi(\lambda)$, where $\lambda \in (\varLambda_{\mathrm{ico}}(t,W)\cap(t+\alpha+H^{(\tau,0,1)}))-t-\alpha$. It follows that $\lambda \in \mathrm{Im}[\mathbb{I}] \cap H^{(\tau,0,1)}$ and $(\alpha+\lambda)^\star = \alpha^\star + \lambda^\star \in W$. Lemma 1.99 implies that $\Phi(\lambda) \in \mathcal{O}_5$, say $\Phi(\lambda) = \beta + \gamma\zeta_5$ for suitable $\beta,\gamma \in \mathcal{O}_5$. One can now easily verify that

$$\left(\Phi(\lambda)\right)^{\star_5} = \beta' + \gamma'\zeta_5^3 = \Phi^\star(\lambda^\star) \in W_\alpha\,.$$

Conversely, suppose that $z \in \mathcal{O}_5$ satisfies $z^{\star_5} \in W_\alpha$. Then, there are suitable $\beta,\gamma \in \mathcal{O}_5$ such that $z = \beta + \gamma\zeta_5$ and, consequently, $z^{\star_5} = \beta' + \gamma'\zeta_5^3 \in W_\alpha$. By definition of W_α, one has $z^{\star_5} = \Phi^\star(\lambda)$, where $\lambda \in H^{(\tau',0,1)}$ satisfies $\alpha^\star + \lambda \in W$. Clearly, there are $r,s \in \mathbb{R}$ such that $\lambda = r(0,1,0)^t + s\frac{1}{2}(-1,-\tau,\tau')^t$, whence $\Phi^\star(\lambda) = r + s\zeta_5^3$. The linear independence of 1 and ζ_5^3 over $\mathbb{R}$ now implies that $r = \beta$ and $s = \gamma$, so that $\lambda \in [\mathrm{Im}[\mathbb{I}]]^\star$. Moreover, denoting by $.^{-\star}$ the inverse of the co-restriction $.^\star : \mathrm{Im}[\mathbb{I}] \longrightarrow [\mathrm{Im}[\mathbb{I}]]^\star$ of $.^\star$ to its image, one has $\lambda^{-\star} \in (\varLambda_{\mathrm{ico}}(t,W) \cap t+\alpha+H^{(\tau,0,1)}) - t - \alpha$ and $\Phi(\lambda^{-\star}) = \beta + \gamma\zeta_5 = z$. This completes the proof. □

COROLLARY 1.102. *Let* $\varLambda_{\mathrm{ico}}(t,W)$ *be an icosahedral model set. Then, for every* $\alpha \in \mathrm{Im}[\mathbb{I}]$*, the set*

$$\Phi\left[\left(\varLambda_{\mathrm{ico}}(t,W) \,\cap\, \left(t+\alpha+H^{(\tau,0,1)}\right)\right) - t - \alpha\right] \tag{1.29}$$

is a subset of a cyclotomic model set.

PROOF. This follows immediately from Proposition 1.101 in conjunction with the definition of cyclotomic model sets; see Definition 1.73. □

REMARK 1.103. In the case of *generic* icosahedral model sets with, e.g., polytopal or ball-shaped windows, the set in (1.29) is either empty or a cyclotomic model set with window W_α (in this situation, W_α is indeed a window) as defined in (1.28); cf. Proposition 1.101.

Since we want to use our later results on discrete tomography of cyclotomic model sets, Proposition 1.101 and Corollary 1.102 suggest the following definition.

DEFINITION 1.104. The $\mathrm{Im}[\mathbb{I}]$-directions in $\mathbb{S}^2 \cap H^{(\tau,0,1)}$ are called $\mathrm{Im}[\mathbb{I}]^{(\tau,0,1)}$-*directions.*

We shall now establish an important result on a relation between icosahedral model sets and their underlying module $\mathrm{Im}[\mathbb{I}]$.

DEFINITION 1.105. We denote by m_τ the $\mathcal{O}_5$-module endomorphism of $\mathrm{Im}[\mathbb{I}]$, given by multiplication by τ, i.e., $\alpha \longmapsto \tau\alpha$. Furthermore, we denote by $(m_\tau)^\star$ the $\mathcal{O}_5$-module endomorphism of $[\mathrm{Im}[\mathbb{I}]]^\star$, given by $\alpha^\star \longmapsto (\tau\alpha)^\star$.

LEMMA 1.106. *The map $(m_\tau)^\star$ is contractive with contraction constant $\frac{1}{\tau} \in (0,1)$, i.e., the equality $\|(m_\tau)^\star(\alpha^\star)\| = \frac{1}{\tau}\|\alpha^\star\|$ holds for all $\alpha \in \mathrm{Im}[\mathbb{I}]$.*

PROOF. Let $\alpha \in \mathrm{Im}[\mathbb{I}]$. Observe that $\|(m_\tau)^\star(\alpha^\star)\| = \|(\tau\alpha)^\star\| = \|\tau'\alpha^\star\| = \frac{1}{\tau}\|\alpha^\star\|$. The observation $\frac{1}{\tau} = -1+\tau \in (0,1)$ completes the proof. □

REMARK 1.107. In the above lemma, we used the fact that τ is a PV-number of (full) degree 2 in $\mathcal{O}_5$. In fact, since $\tau' = -1/\tau = 1-\tau$, the PV-number τ is even a PV-unit. Consequently, the $\mathcal{O}_5$-module endomorphisms m_τ and $(m_\tau)^\star$ are actually $\mathcal{O}_5$-module automorphisms.

LEMMA 1.108. *Let $t \in \mathbb{R}^3$ and let $\Lambda_{\mathrm{ico}}(t,W)$ be an icosahedral model set. Then, for any finite set $F \subset \mathrm{Im}[\mathbb{I}]$, there is an expansive homothety $h\colon \mathbb{R}^3 \longrightarrow \mathbb{R}^3$ such that $h[F] \subset \Lambda_{\mathrm{ico}}(t,W)$. In particular, if $t=0$ and $0 \in \mathrm{int}(W)$, there is even a dilatation $d\colon \mathbb{R}^3 \longrightarrow \mathbb{R}^3$ such that $d[F] \subset \Lambda_{\mathrm{ico}}(t,W)$.*

PROOF. From $\mathrm{int}(W) \neq \varnothing$ and the denseness of $[\mathrm{Im}[\mathbb{I}]]^\star$ in $\mathbb{R}^3$, one gets the existence of a suitable $\alpha_0 \in \mathrm{Im}[\mathbb{I}]$ with ${\alpha_0}^\star \in \mathrm{int}(W)$. Consider the open neighbourhood $V := \mathrm{int}(W) - {\alpha_0}^\star$ of 0 in $\mathbb{R}^3$. Next, consider the map $(m_\tau)^\star\colon [\mathrm{Im}[\mathbb{I}]]^\star \longrightarrow [\mathrm{Im}[\mathbb{I}]]^\star$. Since, by Lemma 1.106, the map $(m_\tau)^\star$ is contractive (in the sense which was made precise in that lemma), the existence of a suitable $k \in \mathbb{N}$ is implied such that

$$\big((m_\tau)^\star\big)^k\big[\,[F]^\star\big] \subset V\,.$$

Hence, one has

$$\big\{(\tau^k\alpha+\alpha_0)^\star \,|\, \alpha \in F\big\} \subset \mathrm{int}(W) \subset W$$

and, further, $h[F] \subset \Lambda_{\mathrm{ico}}(t,W)$, where $h\colon \mathbb{R}^3 \longrightarrow \mathbb{R}^3$ is the expansive homothety given by

$$x \longmapsto \tau^k x + (\alpha_0 + t)\,.$$

The additional statement follows immediately from the foregoing proof in connection with the trivial observation that $0 \in \mathrm{Im}[\mathbb{I}]$ maps, under the star map $.^\star$, to $0 \in \mathbb{R}^3$. This completes the proof. □

Note again that, similar to Lemma 1.81, Lemma 1.108 shows that any icosahedral model set $\varLambda_{\text{ico}}(t, W)$ contains suitably scaled versions of any finite subset of its underlying $\mathbb{Z}$-module $\text{Im}[\mathbb{I}]$.

As an easy application of Lemma 1.108, one obtains the following result on the set of $\varLambda_{\text{ico}}(t, W)$-directions (resp., 1-dimensional $\varLambda_{\text{ico}}(t, W)$-subspaces) for icosahedral model sets $\varLambda_{\text{ico}}(t, W)$.

LEMMA 1.109. *Let $\varLambda_{\text{ico}}(t, W)$ be an icosahedral model set. Then, one has:*

(a) *The set of $\varLambda_{\text{ico}}(t, W)$-directions is precisely the set of* $\text{Im}[\mathbb{I}]$*-directions.*
(b) *The set of* 1*-dimensional $\varLambda_{\text{ico}}(t, W)$-subspaces is precisely the set of* 1*-dimensional* $\text{Im}[\mathbb{I}]$*-subspaces.*

PROOF. Let us start with (a). Since one has $\varLambda_{\text{ico}}(t, W) - \varLambda_{\text{ico}}(t, W) \subset \text{Im}[\mathbb{I}]$, every $\varLambda_{\text{ico}}(t, W)$-direction is an $\text{Im}[\mathbb{I}]$-direction. For the converse, let $u \in \mathbb{S}^2$ be an $\text{Im}[\mathbb{I}]$-direction, say parallel to $\alpha \in \text{Im}[\mathbb{I}] \setminus \{0\}$. By Lemma 1.108, there is a homothety $h\colon \mathbb{R}^3 \longrightarrow \mathbb{R}^3$ such that $h[\{0, \alpha\}] \subset \varLambda_{\text{ico}}(t, W)$. It follows that $h(\alpha) - h(0) \in (\varLambda_{\text{ico}}(t, W) - \varLambda_{\text{ico}}(t, W)) \setminus \{0\}$. Since $h(\alpha) - h(0)$ is parallel to α, the assertion follows. Part (b) follows from similar arguments. □

1.3. Discrete tomography of model sets

1.3.1. X-rays.

DEFINITION 1.110. Let $d \in \mathbb{N}$ and let $F \in \mathcal{F}(\mathbb{R}^d)$. Furthermore, let $u \in \mathbb{S}^{d-1}$ be a direction and let $\mathcal{L}_u^d$ be the set of lines in direction u in $\mathbb{R}^d$. Then, the (*discrete parallel*) *X-ray* of F *in direction* u is the function $X_u F : \mathcal{L}_u^d \longrightarrow \mathbb{N}_0 := \mathbb{N} \cup \{0\}$, defined by

$$X_u F(\ell) := \text{card}(F \cap \ell) = \sum_{x \in \ell} \mathbb{1}_F(x)\,.$$

Moreover, the *support* $(X_u F)^{-1}[\mathbb{N}]$ of $X_u F$, i.e., the set of lines in $\mathcal{L}_u^d$ which pass through at least one point of F, is denoted by $\text{supp}(X_u F)$. For $z \in \mathbb{R}^d$, we denote by ℓ_u^z the element of $\mathcal{L}_u^d$ which passes through z. Moreover, for $S \subset \mathbb{R}^d$, we denote by $\mathcal{L}_u^S$ the subset of $\mathcal{L}_u^d$ consisting of all elements of the form ℓ_u^z, where $z \in S$, i.e., lines in $\mathcal{L}_u^d$ which pass through at least one point of S.

REMARK 1.111. In the situation of Definition 1.110, $\text{supp}(X_u F)$ is finite and, moreover, the cardinality of F is implicit in the X-ray, since one has

$$\sum_{\ell \in \text{supp}(X_u F)} X_u F(\ell) = \text{card}(F)\,.$$

LEMMA 1.112. *Let $d \in \mathbb{N}$ and let $u \in \mathbb{S}^{d-1}$ be a direction. If $F, F' \in \mathcal{F}(\mathbb{R}^d)$, one has:*

(a) $X_u F = X_u F'$ *implies* $\text{card}(F) = \text{card}(F')$.
(b) *If $X_u F = X_u F'$, the centroids of F and F' lie on the same line parallel to u.*

PROOF. See [**33**, Lemma 5.1 and Lemma 5.4]. □

The following property is straight-forward.

LEMMA 1.113. *Let $h : \mathbb{R}^2 \longrightarrow \mathbb{R}^2$ be a homothety and let $U \subset \mathbb{S}^1$ be a finite set of directions. Then, one has:*

(a) *If P is a U-polygon (recall Definition 1.2(e)), the image $h[P]$ of P under h is again a U-polygon.*

(b) *If F_1 and F_2 are elements of $\mathcal{F}(\mathbb{R}^2)$ with the same X-rays in the directions of U, the sets $h[F_1]$ and $h[F_2]$ also have the same X-rays in the directions of U.* □

1.3.2. (Successive) determination by X-rays or projections.

DEFINITION 1.114. Let $d \geq 2$, let $\mathcal{E} \subset \mathcal{F}(\mathbb{R}^d)$, and let $m \in \mathbb{N}$. Further, let U be a finite set of directions and let $\mathcal{T}$ be a finite set of 1-dimensional subspaces T of $\mathbb{R}^d$.

(a) We say that $\mathcal{E}$ is *determined* by the X-rays in the directions of U if, for all $F, F' \in \mathcal{E}$, one has

$$(X_u F = X_u F' \ \forall u \in U) \implies F = F' .$$

(b) We say that $\mathcal{E}$ is *determined* by the (orthogonal) projections on the orthogonal complements $T^{\perp}$ of the subspaces T of $\mathcal{T}$ if, for all $F, F' \in \mathcal{E}$, one has

$$(F|T^{\perp} = F'|T^{\perp} \ \forall T \in \mathcal{T}) \implies F = F' .$$

(c) We say that $\mathcal{E}$ is *successively determined* by the X-rays in the directions of U, say $U = \{u_1, \dots, u_m\}$, if, for a given $F \in \mathcal{E}$, these can be chosen inductively (i.e., the choice of u_j depending on all $X_{u_k} F$ with $k \in \{1, \dots, j-1\}$) such that, for all $F' \in \mathcal{E}$, one has

$$(X_u F' = X_u F \ \forall u \in U) \implies F' = F .$$

(d) We say that $\mathcal{E}$ is *successively determined* by the (orthogonal) projections on the orthogonal complements $T^{\perp}$ of the subspaces T of $\mathcal{T}$, say $\mathcal{T} = \{T_1, \dots, T_m\}$, if, for a given $F \in \mathcal{E}$, these can be chosen inductively (i.e., the choice of T_j depending on all $F|T_k^{\perp}$ with $k \in \{1, \dots, j-1\}$) such that, for all $F' \in \mathcal{E}$, one has

$$(F|T^{\perp} = F'|T^{\perp} \ \forall T \in \mathcal{T}) \implies F = F' .$$

(e) We say that $\mathcal{E}$ is *determined* (resp., *successively determined*) by m X-rays if there is a set U of m pairwise non-parallel directions such that $\mathcal{E}$ is determined (resp., successively determined) by the X-rays in the directions of U.

(f) We say that $\mathcal{E}$ is *determined* (resp., *successively determined*) by m (orthogonal) projections on orthogonal complements of 1-dimensional subspaces if there is a set $\mathcal{T}$ of m pairwise non-parallel 1-dimensional subspaces of $\mathbb{R}^d$ such that $\mathcal{E}$ is determined (resp., successively determined) by the (orthogonal) projections on the orthogonal complements $T^{\perp}$ of the subspaces T of $\mathcal{T}$.

REMARK 1.115. Let $\mathcal{E} \subset \mathcal{F}(\mathbb{R}^d)$. Note that if $\mathcal{E}$ is determined by a set of X-rays (resp., projections), then $\mathcal{E}$ is also successively determined by the same X-rays (resp., projections).

1.3.3. General setting. In order to define the analogue of a specific crystal in the case of aperiodic model sets, one additional difficulty, in comparison to the crystallographic case, stems from the fact that it is not sufficient to consider one pattern and its translates to define the setting. Hence, to define the analogue of a specific crystal, one has to add all infinite patterns that emerge as limits of sequences of translates defined in the local topology (LT). Here, two patterns are ε-close if, after a translation by a distance of at most ε, they agree on a ball of radius $1/\varepsilon$ around the origin. If the starting pattern P is crystallographic, no new patterns are added; but if P is a generic aperiodic model set, one ends up with uncountably many different patterns, even up to translations! Nevertheless, all of them are locally indistinguishable (LI). This means that every *finite* patch in Λ also appears in any of the other elements of the LI-class and vice versa; see [**5**] for details. The restriction to the *generic* case is the proper analogue of the restriction to *perfect* lattices and their translates in the classical setting. The entire LI-class $\mathrm{LI}(\Lambda(t,W))$ of a regular, generic model set $\Lambda(t,W)$ can be shown to consist of all sets $t' + \Lambda(\tau + W)$, with $t' \in \mathbb{R}^d$ and $\tau \in H$ such that $\mathrm{bd}(\tau + W) \cap [L]^\star = \varnothing$ (i.e., τ is in a generic position), and all patterns obtained as limits of sequences $t' + \Lambda(\tau_n + W)$, with all τ_n in a generic position; see [**5, 60**]. Each such limit is then a *subset* of some $t + \Lambda(\tau + W)$, as $\mathrm{cl}(W)$ was assumed compact, but τ might not be in a generic position. Note that translates $\tau + W$, $\tau \in H$, of the window W in the internal space H are windows as well. In view of the latter, for the discrete tomography of aperiodic model sets, we must make sure that we deal with finite subsets of elements of the LI-class of fixed *generic* model sets $\Lambda(t,W)$, i.e., finite sets whose image under the star map lies in the *interior* $\mathrm{int}(W)$ of the window. The general setting for the uniqueness problem of discrete tomography of model sets looks as follows. Let $\Lambda \subset \mathbb{R}^d$, $d \geq 2$, be a regular, generic model set. Then, for the uniqueness problems, we are interested in the (successive) determination of the set

$$\bigcup_{\Lambda' \in \mathrm{LI}(\Lambda)} \mathcal{F}(\Lambda')$$

or suitable subsets thereof by the X-rays in a small number of Λ-directions. For the algorithmic problems, we are mainly interested in the reconstruction of finite subsets of sets $\Lambda' \in \mathrm{LI}(\Lambda)$ from their X-rays in a small number of Λ-directions. More precisely, the main algorithmic problems of discrete tomography of model sets are the following.

DEFINITION 1.116 (Consistency, Reconstruction, and Uniqueness Problem). Let $\Lambda \subset \mathbb{R}^d$, $d \geq 2$, be a regular, generic model set with underlying $\mathbb{Z}$-module L. Further, let $u_1, \dots, u_m \in \mathbb{S}^{d-1}$ be $m \geq 2$ pairwise non-parallel Λ-directions. The corresponding consistency, reconstruction and uniqueness problems are defined as follows.

CONSISTENCY.
Given functions $p_{u_j} : \mathcal{L}^d_{u_j} \longrightarrow \mathbb{N}_0$, $j \in \{1, \dots, m\}$, whose supports are finite and satisfy $\mathrm{supp}(p_{u_j}) \subset \mathcal{L}^L_{u_j}$, decide whether there is a finite set F which is contained in a set $\Lambda' \in \mathrm{LI}(\Lambda)$ and satisfies $X_{u_j}F = p_{u_j}$, $j \in \{1, \dots, m\}$.

RECONSTRUCTION.
Given functions $p_{u_j} : \mathcal{L}^d_{u_j} \longrightarrow \mathbb{N}_0$, $j \in \{1, \dots, m\}$, whose supports are finite and satisfy $\mathrm{supp}(p_{u_j}) \subset \mathcal{L}^L_{u_j}$, decide whether there exists a finite subset F which is contained in a set $\Lambda' \in \mathrm{LI}(\Lambda)$ and satisfies $X_{u_j}F = p_{u_j}$, $j \in \{1, \dots, m\}$, and, if so, construct one such F.

UNIQUENESS.
Given a finite subset F of a set $\Lambda' \in \mathrm{LI}(\Lambda)$, decide whether there is a different finite set F' that is also a subset of a set $\Lambda'' \in \mathrm{LI}(\Lambda)$ and satisfies $X_{u_j}F = X_{u_j}F'$, $j \in \{1, \dots, m\}$.

REMARK 1.117. The above setting for the discrete tomography of model sets is motivated by the practice of quantitative HRTEM. This is due to the fact that, because of the symmetries of genuine (quasi)crystals (e.g., icosahedral symmetry) in conjunction with their pure point diffractiveness, the determination of the rotational orientation of a (quasi)crystalline probe in an electron microscope can rather easily be done in the diffraction mode, prior to taking images in the high-resolution mode, though, in general, a natural choice of a translational origin is *not* possible. Therefore, in order to prove practically relevant and rigorous results, one has to deal with the 'non-anchored' case of the whole LI-class of a regular, generic model set Λ, rather than dealing with the 'anchored' case of a fixed such Λ. It will turn out in the following that the treatment of the 'non-anchored' case is often feasible. Moreover, in Section 2.3, we shall even be able to provide positive uniqueness results for which we can make sure that all the Λ-directions used correspond to dense lines in the corresponding model sets, the latter meaning that the resolution coming from these directions is rather high. Hence, we believe that these results look promising for real applications.

1.3.4. Grids.

1.3.4.1. *General setting.*

DEFINITION 1.118. Let $d \geq 2$ and let $U \subset \mathbb{S}^{d-1}$ be a finite set of pairwise non-parallel directions.

(a) Let $p_u : \mathcal{L}_u^d \longrightarrow \mathbb{N}_0$, $u \in U$, be functions whose supports $\mathrm{supp}(p_u) = p_u^{-1}[\mathbb{N}]$ are finite. Then, the *grid* $G_{\{p_u|u \in U\}}$ of the set $\{p_u | u \in U\}$ is defined by

$$G_{\{p_u|u\in U\}} := \bigcap_{u\in U} \left(\bigcup_{\ell \in \mathrm{supp}(p_u)} \ell \right) \subset \mathbb{R}^d .$$

(b) Let F be a finite subset of $\mathbb{R}^d$. We define the *grid* G_U^F of F with respect to the X-rays in the directions of U as

$$G_U^F := \bigcap_{u\in U} \left(\bigcup_{\ell \in \mathrm{supp}(X_u F)} \ell \right) \subset \mathbb{R}^d .$$

(c) Let S be a subgroup of $\mathbb{R}^d$. We define the *complete grid* G_U^S of S with respect to U as

$$G_U^S := \bigcap_{u\in U} \left(\bigcup_{\ell \in \mathcal{L}_u^S} \ell \right) \subset \mathbb{R}^d .$$

REMARK 1.119. Note that, in the situation of Definition 1.118, one has the inclusion

$$S \subset G_U^S .$$

Note further that the complete grid G_U^S of S with respect to U is a subgroup of $\mathbb{R}^d$.

LEMMA 1.120. *Let $d \geq 2$ and let $U \subset \mathbb{S}^{d-1}$ be a finite set of pairwise non-parallel directions. Then, for all finite subsets F, F' of $\mathbb{R}^d$, one has*

$$(X_u F = X_u F' \ \forall u \in U) \implies F, F' \subset G_U^F = G_U^{F'} .$$

PROOF. This follows immediately from the definition of grids. □

We need the following definition.

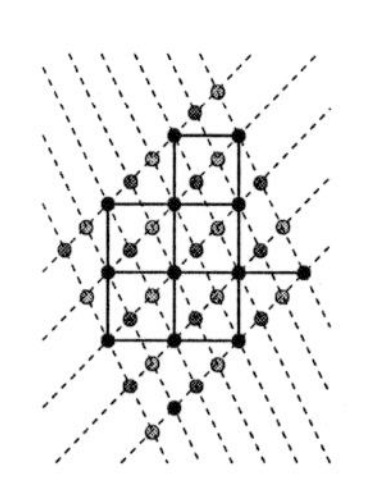

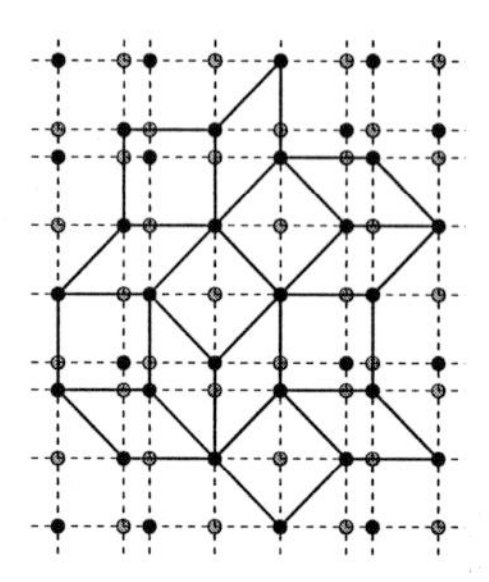

FIGURE 1.6. Grids arising from two $\mathcal{O}_n$-directions: On the left, the grid generated by the X-rays of a finite subset of a translate of $\mathcal{O}_4 = \mathbb{Z}^2$ in the two non-parallel $\mathcal{O}_4$-directions u_{1+i} and u_{1-2i}. The three equivalence classes modulo $\mathcal{O}_4$ are marked by different greyscales. On the right, the grid generated by the X-rays of a finite subset of a translate of $\varLambda_{\mathrm{AB}}$ in the two non-parallel $\mathcal{O}_8$-directions 1 and $\zeta_8^2 = i$. Two equivalence classes modulo $\mathcal{O}_8$ are present, which are again marked by different greyscales.

DEFINITION 1.121. For $d \in \mathbb{N}$ and a subgroup L of $\mathbb{R}^d$, define the equivalence relation $\sim_L$ on $\mathbb{R}^d$ by

$$x \sim_L y \;:\Longleftrightarrow\; x - y \in L\,.$$

Further, if $x, y \in \mathbb{R}^d$ satisfy $x \sim_L y$, we say that x and y are *equivalent modulo L*.

1.3.4.2. *Grids generated by finite subsets of rings of cyclotomic integers.* For the discrete tomography of cyclotomic model sets, we are naturally interested in the following special situation. Let $n \in \mathbb{N} \setminus \{1,2\}$, let $U \subset \mathbb{S}^1$ be a finite set of pairwise non-parallel $\mathcal{O}_n$-directions, and let F be a finite subset of $t + \mathcal{O}_n$, where $t \in \mathbb{R}^2$. Clearly, the grid G_U^F of F with respect to the X-rays in the directions of U may be a proper superset of F. In fact, it may even contain points that lie in a *different* translate of $\mathcal{O}_n$ than F itself; see Figure 1.6.

Next, we will show that the number of equivalence classes of such a grid G_U^F modulo $\mathcal{O}_n$ is uniformly bounded by a number that depends on the given $\mathcal{O}_n$-directions but is independent of the X-ray data (Proposition 1.125 below). Further, since we also want to prove uniqueness results, we are interested in the case where the phenomenon of multiple equivalence classes modulo $\mathcal{O}_n$ in grids of the form G_U^F cannot occur. It will be shown (Theorem 1.130 below) that this goal can be reached by allowing only a special kind of sets U of at least two $\mathcal{O}_n$-directions.

PROPOSITION 1.122. *Let $n \in \mathbb{N}\setminus\{1,2\}$ and let $u, u' \in \mathbb{S}^1$ be two non-parallel $\mathcal{O}_n$-directions, say $u = u_o$ and $u' = u_{o'}$, where $o, o' \in \mathcal{O}_n \setminus \{0\}$ are non-parallel. Then, the complete grid $G_{\{u_o,u_{o'}\}}^{\mathcal{O}_n}$ of $\mathcal{O}_n$ with respect to $\{u_o, u_{o'}\}$ satisfies*

$$\mathcal{O}_n \subset G_{\{u_o,u_{o'}\}}^{\mathcal{O}_n} \subset \mathcal{M}_{\{o,o'\}} \subset \mathbb{K}_n = \mathbb{Q}\mathcal{O}_n \subset \mathbb{C},$$

where

$$\mathcal{M}_{\{o,o'\}} := \big\langle \{o/(\alpha_o\beta_{o'} - \beta_o\alpha_{o'}), o'/(\alpha_o\beta_{o'} - \beta_o\alpha_{o'})\} \big\rangle_{\mathcal{O}_n}$$

and $\alpha_o, \alpha_{o'}, \beta_o, \beta_{o'} \in \mathcal{O}_n$ *are uniquely determined by* $o = \alpha_o + \beta_o\zeta_n$ *and* $o' = \alpha_{o'} + \beta_{o'}\zeta_n$ *(cf. Lemma 1.9(a)). Further,* $\langle\,.\,\rangle_{\mathcal{O}_n}$ *denotes the* $\mathcal{O}_n$*-linear hull.*

PROOF. First, note that the linear independence of $\{o, o'\}$ and $\{1, \zeta_n\}$ over $\mathbb{R}$ implies that $\alpha_o\beta_{o'} - \beta_o\alpha_{o'} \neq 0$. For the first inclusion, see Remark 1.119. The third and the last inclusion are obvious, whereas the equality $\mathbb{K}_n = \mathbb{Q}\mathcal{O}_n$ follows from Proposition 1.11 in conjunction with Remark 1.24. Next, we claim that $\mathcal{O}_n \subset \mathcal{M}_{\{o,o'\}}$. Let $z \in \mathcal{O}_n$. By Lemma 1.9(a), there are unique $\varphi, \psi \in \mathcal{O}_n$ with $z = \varphi + \psi\zeta_n$. By the linear independence of $\{o, o'\}$ over $\mathbb{R}$, there are unique $x, y \in \mathbb{R}$ with $xo + yo' = z$. Hence

$$(x\alpha_o + y\alpha_{o'} - \varphi) + (x\beta_o + y\beta_{o'} - \psi)\zeta_n = 0$$

and, using the linear independence of $\{1, \zeta_n\}$ over $\mathbb{R}$, we get that

$$x\alpha_o + y\alpha_{o'} - \varphi = x\beta_o + y\beta_{o'} - \psi = 0\,.$$

In matrix notation, this means that

$$\begin{pmatrix} \alpha_o & \alpha_{o'} \\ \beta_o & \beta_{o'} \end{pmatrix} \begin{pmatrix} x \\ y \end{pmatrix} = \begin{pmatrix} \varphi \\ \psi \end{pmatrix}.$$

Cramer's rule now implies that

$$x = (\varphi\beta_{o'} - \psi\alpha_{o'})/(\alpha_o\beta_{o'} - \beta_o\alpha_{o'}) \in \mathcal{O}_n/(\alpha_o\beta_{o'} - \beta_o\alpha_{o'})$$

and

$$y = (\alpha_o\psi - \beta_o\varphi)/(\alpha_o\beta_{o'} - \beta_o\alpha_{o'}) \in \mathcal{O}_n/(\alpha_o\beta_{o'} - \beta_o\alpha_{o'})\,.$$

This proves our claim.

Finally, consider $g \in G^{\mathcal{O}_n}_{\{u_o, u_{o'}\}}$. By definition, there are elements $v, w \in \mathcal{O}_n$ such that $\{g\} = (v + \mathbb{R}o) \cap (w + \mathbb{R}o')$. Moreover, there are unique $x, y \in \mathbb{R}$ with $g = v + xo = w + yo'$. Hence, $xo + (-y)o' = w - v \in \mathcal{O}_n$ and, by the same calculation as above, we get that $x, y \in \mathcal{O}_n/(\alpha_o\beta_{o'} - \beta_o\alpha_{o'})$. Together with our first claim, this shows that $g \in \mathcal{M}_{\{o,o'\}}$. □

REMARK 1.123. In the situation of Proposition 1.122, note that $\mathcal{M}_{\{o,o'\}}$ is an $\mathcal{O}_n$-module of rank 2 with basis

$$\{o/(\alpha_o\beta_{o'} - \beta_o\alpha_{o'}), o'/(\alpha_o\beta_{o'} - \beta_o\alpha_{o'})\}\,.$$

Observe further that, for $n \in \mathbb{N} \setminus \{1, 2, 3, 4, 6\}$, the complete grid $G^{\mathcal{O}_n}_{\{u_o, u_{o'}\}}$ is a dense subset of the plane, because already its subset $\mathcal{O}_n$ has this property; cf. Remark 1.10 and Remark 1.119.

LEMMA 1.124. *In the situation of Proposition 1.122,* $\mathcal{M}_{\{o,o'\}}$ *is a* $\mathbb{Z}$*-module of rank* $\phi(n)$.

PROOF. This is an immediate consequence of the facts that $\mathcal{M}_{\{o,o'\}}$ is an $\mathcal{O}_n$-module of rank 2 and $\mathcal{O}_n$ is a $\mathbb{Z}$-module of rank $\phi(n)/2$; see Remark 1.123 and Remark 1.24. □

The following lemma shows that $\mathcal{M}_{\{o,o'\}}$, and thus $G^{\mathcal{O}_n}_{\{u_o, u_{o'}\}}$, decomposes into finitely many equivalence classes modulo $\mathcal{O}_n$, the number of equivalence classes only depending on $\{o, o'\}$.

PROPOSITION 1.125. *In the situation of Proposition 1.122, the subgroup index* $[\mathcal{M}_{\{o,o'\}} : \mathcal{O}_n]$ *is finite. Hence, there are* $c \in \mathbb{N}$ *and* $t_1, t_2, \dots, t_c \in \mathcal{M}_{\{o,o'\}}$ *such that*

$$\mathcal{M}_{\{o,o'\}} = \dot{\bigcup}_{j=1}^{c} (t_j + \mathcal{O}_n)\,,$$

where, without restriction, $t_1 = 0$. *It follows that every subset* G *of* $\mathcal{M}_{\{o,o'\}}$ *satisfies the decomposition*

$$G = \dot{\bigcup}_{j=1}^{c} (G \cap (t_j + \mathcal{O}_n)) .$$

PROOF. By Lemma 1.124, $\mathcal{M}_{\{o,o'\}}$ is a $\mathbb{Z}$-module of rank $\phi(n)$. Moreover, $\mathcal{M}_{\{o,o'\}}$ is obviously torsion-free. But $\mathcal{O}_n$ is a $\mathbb{Z}$-module of rank $\phi(n)$ as well; see Remark 1.24. Now, Proposition 1.4 yields the result. □

DEFINITION 1.126. For elements $o \in \mathcal{O}_n \setminus \{0\}$ and $o' \in \mathcal{O}_n$, we say that o *divides* o' and write $o|o'$ if $\frac{o'}{o} \in \mathcal{O}_n$.

DEFINITION 1.127. Let $\mathbb{K}/\mathbb{k}$ be an extension of algebraic number fields, say of degree $d := [\mathbb{K} : \mathbb{k}] \in \mathbb{N}$. Further, let $\mathcal{O}_{\mathbb{K}}$ (resp., $o_{\mathbb{k}}$) be the ring of integers of $\mathbb{K}$ (resp., $\mathbb{k}$). Then, a subset $\{o_1, \dots, o_d\} \subset \mathcal{O}_{\mathbb{K}}$ is called a *relative integral basis* of $\mathbb{K}/\mathbb{k}$ if it is an $o_{\mathbb{k}}$-basis of the $o_{\mathbb{k}}$-module $\mathcal{O}_{\mathbb{K}}$, i.e., if every element $o \in \mathcal{O}_{\mathbb{K}}$ is uniquely expressible as an $o_{\mathbb{k}}$-linear combination of $\{o_1, \dots, o_d\}$.

REMARK 1.128. By Lemma 1.9(b), for $n \in \mathbb{N} \setminus \{1, 2\}$, one has $[\mathbb{K}_n : \mathbb{k}_n] = 2$. Moreover, by Proposition 1.23, one has the identities $\mathcal{O}_{\mathbb{K}_n} = \mathcal{O}_n$ and $o_{\mathbb{k}_n} = o_n$. Let o, o' be two non-parallel elements of $\mathcal{O}_n$. Then, $\{o, o'\}$ is a relative integral basis of $\mathbb{K}_n/\mathbb{k}_n$ if and only if the o_n-linear hull $\langle\{o, o'\}\rangle_{o_n}$ of $\{o, o'\}$ equals $\mathcal{O}_n$.

PROPOSITION 1.129. *Let* $n \in \mathbb{N} \setminus \{1, 2\}$ *and let* o, o' *be two non-parallel elements of* $\mathcal{O}_n$, *say* $o = \alpha_o + \beta_o \zeta_n$ *and* $o' = \alpha_{o'} + \beta_{o'} \zeta_n$ *for uniquely determined* $\alpha_o, \alpha_{o'}, \beta_o, \beta_{o'} \in o_n$ *(cf. Lemma 1.9(a)). The following statements are equivalent:*

(i) $\alpha_o \beta_{o'} - \beta_o \alpha_{o'} \in o_n^{\times}$.

(ii) $\alpha_o \beta_{o'} - \beta_o \alpha_{o'} | o$ *and* $\alpha_o \beta_{o'} - \beta_o \alpha_{o'} | o'$.

(iii) $\{o, o'\}$ *is a relative integral basis of* $\mathbb{K}_n/\mathbb{k}_n$.

Moreover, each of the above conditions (i)-(iii) implies the equation

$$G^{\mathcal{O}_n}_{\{u_o, u_{o'}\}} = \mathcal{O}_n .$$

PROOF. Direction (i) ⇒ (ii) is immediate. Next, we show direction (ii) ⇒ (i). Suppose that $\alpha_o \beta_{o'} - \beta_o \alpha_{o'} | o$ and $\alpha_o \beta_{o'} - \beta_o \alpha_{o'} | o'$, say $(\alpha_o \beta_{o'} - \beta_o \alpha_{o'})(\gamma + \delta \zeta_n) = o$ and $(\alpha_o \beta_{o'} - \beta_o \alpha_{o'})(\gamma' + \delta' \zeta_n) = o'$ for suitable $\gamma, \gamma', \delta, \delta' \in o_n$. Using the $\mathbb{R}$-linear independence of $\{1, \zeta_n\}$, the latter implies the equations

$$\begin{aligned}
(\alpha_o \beta_{o'} - \beta_o \alpha_{o'})\gamma &= \alpha_o , \\
(\alpha_o \beta_{o'} - \beta_o \alpha_{o'})\delta &= \beta_o , \\
(\alpha_o \beta_{o'} - \beta_o \alpha_{o'})\gamma' &= \alpha_{o'} , \\
(\alpha_o \beta_{o'} - \beta_o \alpha_{o'})\delta' &= \beta_{o'} .
\end{aligned}$$

Consequently, one has $\alpha_o \beta_{o'} - \beta_o \alpha_{o'} = (\alpha_o \beta_{o'} - \beta_o \alpha_{o'})^2 (\gamma \delta' - \delta \gamma')$. Further, dividing by $\alpha_o \beta_{o'} - \beta_o \alpha_{o'}$, one obtains the equation

$$1 = (\alpha_o \beta_{o'} - \beta_o \alpha_{o'}) (\gamma \delta' - \delta \gamma') ,$$

and the assertion follows. For direction (i) $\Rightarrow$ (iii), first observe that Proposition 1.122 particularly shows that

$$\mathcal{O}_n \subset \big\langle \{o/(\alpha_o\beta_{o'} - \beta_o\alpha_{o'}), o'/(\alpha_o\beta_{o'} - \beta_o\alpha_{o'})\} \big\rangle_{\mathcal{O}_n} \subset \mathcal{O}_n\,,$$

hence

$$\langle\{o/(\alpha_o\beta_{o'} - \beta_o\alpha_{o'}), o'/(\alpha_o\beta_{o'} - \beta_o\alpha_{o'})\}\rangle_{\mathcal{O}_n} = \langle\{o,o'\}\rangle_{\mathcal{O}_n} = \mathcal{O}_n\,.$$

By Remark 1.128, the assertion follows. Finally, let us prove direction (iii) $\Rightarrow$ (i). Here, by Remark 1.128, there are $\gamma,\gamma',\delta,\delta' \in \mathcal{O}_n$ such that $\gamma o + \delta o' = 1$ and $\gamma' o + \delta' o' = \zeta_n$. Using the $\mathbb{R}$-linear independence of $\{1,\zeta_n\}$, the latter implies that $\gamma\alpha_o + \delta\alpha_{o'} = 1$, $\gamma\beta_o + \delta\beta_{o'} = 0$, $\gamma'\alpha_o + \delta'\alpha_{o'} = 0$ and $\gamma'\beta_o + \delta'\beta_{o'} = 1$. Hence, one has

$$\begin{aligned} 1 &= (\gamma\alpha_o + \delta\alpha_{o'})(\gamma'\beta_o + \delta'\beta_{o'}) - (\gamma\beta_o + \delta\beta_{o'})(\gamma'\alpha_o + \delta'\alpha_{o'}) \\ &= (\alpha_o\beta_{o'} - \beta_o\alpha_{o'})(\gamma\delta' - \gamma'\delta)\,, \end{aligned}$$

and the assertion follows. The additional statement follows immediately from Proposition 1.122, which in this case shows that

$$\mathcal{O}_n \subset G^{\mathcal{O}_n}_{\{u_o,u_{o'}\}} \subset \big\langle \{o/(\alpha_o\beta_{o'} - \beta_o\alpha_{o'}), o'/(\alpha_o\beta_{o'} - \beta_o\alpha_{o'})\} \big\rangle_{\mathcal{O}_n} \subset \mathcal{O}_n\,.$$

This completes the proof. □

THEOREM 1.130. *Let $n \in \mathbb{N}\backslash\{1,2\}$, let $O \subset \mathcal{O}_n\backslash\{0\}$ be a finite set of $m \geq 2$ pairwise non-parallel elements. Suppose the existence of two different elements $o,o' \in O$ satisfying one of the equivalent conditions (i)-(iii) of Proposition 1.129. Then, setting $U_O := \{u_o \,|\, o \in O\} \subset \mathbb{S}^1$, for all $t \in \mathbb{R}^2$ and all finite subsets F of $t + \mathcal{O}_n$, one has the inclusion $G^F_{U_O} \subset t + \mathcal{O}_n$.*

PROOF. Let $t \in \mathbb{R}^2$ and let F be a finite subset of $t + \mathcal{O}_n$. It follows that $F - t$ is a finite subset of $\mathcal{O}_n$. Clearly, one has the inclusion $G^{F-t}_{U_O} \subset G^{\mathcal{O}_n}_{U_O} \subset G^{\mathcal{O}_n}_{\{u_o,u_{o'}\}}$. Hence, by the additional statement of Proposition 1.129, one further obtains

$$G^{F-t}_{U_O} \subset G^{\mathcal{O}_n}_{U_O} \subset G^{\mathcal{O}_n}_{\{u_o,u_{o'}\}} = \mathcal{O}_n\,.$$

The equality $G^{F-t}_{U_O} = G^F_{U_O} - t$ completes the proof. □

1.3.4.3. *Grids generated by finite sets of imaginary parts of icosians.*

PROPOSITION 1.131. *Let $u,u' \in \mathbb{S}^2$ be two non-parallel $\mathrm{Im}[\mathbb{I}]^{(\tau,0,1)}$-directions, say $u = u_o$ and $u' = u_{o'}$, where $o,o' \in \mathrm{Im}[\mathbb{I}] \setminus \{0\}$ are non-parallel. Then, the complete grid $G^{\mathrm{Im}[\mathbb{I}]}_{\{u_o,u_{o'}\}}$ of $\mathrm{Im}[\mathbb{I}]$ with respect to $\{u_o,u_{o'}\}$ satisfies*

$$\mathrm{Im}[\mathbb{I}] \subset G^{\mathrm{Im}[\mathbb{I}]}_{\{u_o,u_{o'}\}} \subset \mathcal{N}_{\{o,o'\}} \subset (\mathbb{k}_5)^3 = \mathbb{Q}\,\mathrm{Im}[\mathbb{I}] \subset \mathbb{R}^3\,,$$

where

$$\mathcal{N}_{\{o,o'\}} := \Big\langle \big\{o/(\alpha_o\beta_{o'} - \beta_o\alpha_{o'}), o'/(\alpha_o\beta_{o'} - \beta_o\alpha_{o'}), \tfrac{1}{2}(1,1,1)^t\big\} \Big\rangle_{\mathcal{O}_5}$$

and $\alpha_o,\alpha_{o'},\beta_o,\beta_{o'} \in \mathcal{O}_5$ are uniquely determined by $\Phi(o) = \alpha_o + \beta_o\zeta_n$ and $\Phi(o') = \alpha_{o'} + \beta_{o'}\zeta_n$ (cf. Lemma 1.9(a)) and Φ is as defined in Definition 1.97 (a). Further, $\langle\,.\,\rangle_{\mathcal{O}_5}$ denotes the $\mathcal{O}_5$-linear hull.

PROOF. First, note that the linear independence of $\{\Phi(o),\Phi(o')\}$ and $\{1,\zeta_n\}$ over $\mathbb{R}$ implies that $\alpha_o\beta_{o'}-\beta_o\alpha_{o'}\neq 0$. For the first inclusion, see Remark 1.119. The third and the last inclusion are obvious, whereas the equality $(\mathbb{k}_5)^3=\mathbb{Q}\,\mathrm{Im}[\mathbb{I}]$ was shown in Corollary 1.100.

Next, we claim that $\mathrm{Im}[\mathbb{I}]\subset\mathcal{N}_{\{o,o'\}}$. Let $x\in\mathrm{Im}[\mathbb{I}]$. The identity $\mathrm{Im}[\mathbb{I}]=\frac{1}{2}\mathcal{M}_{\mathrm{B}}$ in conjunction with Equation (1.24) show that there are uniquely determined $\phi,\psi,\nu\in\mathcal{O}_5$ with

$$x \;=\; \phi\,(0,1,0)^t+\psi\,\tfrac{1}{2}(-1,-\tau',\tau)^t+\nu\tfrac{1}{2}(1,1,1)^t\,.$$

Lemma 1.96 (a) implies that

$$x-\nu\tfrac{1}{2}(1,1,1)^t \;=\; \phi\,(0,1,0)^t+\psi\,\tfrac{1}{2}(-1,-\tau',\tau)^t \;\in\; \mathrm{Im}[\mathbb{I}]\cap H^{(\tau,0,1)}\,,$$

and, further, Proposition 1.122 shows that

$$\Phi\left(x-\nu\tfrac{1}{2}(1,1,1)^t\right) \;=\; \phi+\psi\zeta_5 \;\in\; \mathcal{O}_5 \;\subset\; \mathcal{M}_{\{\Phi(o),\Phi(o')\}}\,,$$

where

$$\mathcal{M}_{\{\Phi(o),\Phi(o')\}} \;:=\; \big\langle\,\{\Phi(o)/(\alpha_o\beta_{o'}-\beta_o\alpha_{o'}),\Phi(o')/(\alpha_o\beta_{o'}-\beta_o\alpha_{o'})\}\,\big\rangle_{\mathcal{O}_5}\,. \tag{1.30}$$

Applying Φ^{-1} now shows that

$$x-\nu\tfrac{1}{2}(1,1,1)^t \;\in\; \big\langle\,\{o/(\alpha_o\beta_{o'}-\beta_o\alpha_{o'}),o'/(\alpha_o\beta_{o'}-\beta_o\alpha_{o'})\}\,\big\rangle_{\mathcal{O}_5}$$

and, further, $x\in\mathcal{N}_{\{o,o'\}}$. This proves that $\mathrm{Im}[\mathbb{I}]\subset\mathcal{N}_{\{o,o'\}}$. Next, consider $g\in G^{\mathrm{Im}[\mathbb{I}]}_{\{u_o,u_{o'}\}}$, say $g=\alpha+\lambda o=\beta+\mu o'$ for suitable $\alpha,\beta\in\mathrm{Im}[\mathbb{I}]$ and $\lambda,\mu\in\mathbb{R}$. It follows that

$$\alpha-\beta \;=\; -\lambda o+\mu o' \;\in\; \mathrm{Im}[\mathbb{I}]\cap H^{(\tau,0,1)}\,,$$

and, by applying Φ and using Proposition 1.122, one obtains

$$\lambda\,\Phi(o) \;=\; \Phi(\alpha-\beta)-\mu\,\Phi(o') \;\in\; G^{\mathrm{Im}[\mathbb{I}]}_{\{u_{\Phi(o)},u_{\Phi(o')}\}} \;\subset\; \mathcal{M}_{\{\Phi(o),\Phi(o')\}}\,,$$

with $\mathcal{M}_{\{\Phi(o),\Phi(o')\}}$ as defined in Equation (1.30). Hence,

$$\begin{aligned}\lambda\,\Phi(o) &= \Phi(\alpha-\beta)-\mu\,\Phi(o')\\ &= \gamma\,\frac{\Phi(o)}{\alpha_o\beta_{o'}-\beta_o\alpha_{o'}}+\delta\,\frac{\Phi(o')}{\alpha_o\beta_{o'}-\beta_o\alpha_{o'}}\end{aligned}$$

for suitable $\gamma,\delta\in\mathcal{O}_5$. In particular, it follows that $\lambda=\frac{\gamma}{\alpha_o\beta_{o'}-\beta_o\alpha_{o'}}$, and, further,

$$g \;=\; \alpha+\lambda o \;=\; \alpha+\frac{\gamma}{\alpha_o\beta_{o'}-\beta_o\alpha_{o'}}o \;\in\; \mathcal{N}_{\{o,o'\}}\,,$$

since $\alpha\in\mathcal{N}_{\{o,o'\}}$ by our first claim. This proves the second inclusion and completes the proof. □

REMARK 1.132. In the situation of Proposition 1.131, note that $\mathcal{N}_{\{o,o'\}}$ is an $\mathcal{O}_5$-module of rank 3 with basis

$$\{o/(\alpha_o\beta_{o'}-\beta_o\alpha_{o'}),o'/(\alpha_o\beta_{o'}-\beta_o\alpha_{o'}),\tfrac{1}{2}(1,1,1)^t\}\,.$$

Further observe that the complete grid $G^{\mathrm{Im}[\mathbb{I}]}_U$ is a dense subset of $\mathbb{R}^3$, because already its subset $\mathrm{Im}[\mathbb{I}]$ has this property; cf. Remark 1.10 and Remark 1.119.

LEMMA 1.133. *In the situation of Proposition 1.131, $\mathcal{N}_{\{o,o'\}}$ is a $\mathbb{Z}$-module of rank* 6.

PROOF. This is an immediate consequence of the facts that $\mathcal{N}_{\{o,o'\}}$ is an $\mathcal{O}_5$-module of rank 3 and $\mathcal{O}_5$ is a $\mathbb{Z}$-module of rank 2; see Proposition 1.131 and Remark 1.24. □

The following lemma shows that $\mathcal{N}_{\{o,o'\}}$, and thus $G^{\mathrm{Im}[\mathbb{I}]}_{\{u_o,u_{o'}\}}$, decomposes into finitely many equivalence classes modulo $\mathrm{Im}[\mathbb{I}]$, the number of equivalence classes only depending on $\{o,o'\}$.

PROPOSITION 1.134. *In the situation of Proposition 1.131, the subgroup index* $[\mathcal{N}_{\{o,o'\}} : \mathrm{Im}[\mathbb{I}]]$ *is finite. Hence, there are* $c \in \mathbb{N}$ *and* $t_1, t_2, \dots, t_c \in \mathcal{N}_{\{o,o'\}}$ *such that*

$$\mathcal{N}_{\{o,o'\}} = \dot{\bigcup}_{j=1}^{c} (t_j + \mathrm{Im}[\mathbb{I}]) ,$$

where, without restriction, $t_1 = 0$. *It follows that every subset* G *of* $\mathcal{N}_{\{o,o'\}}$ *satisfies the decomposition*

$$G = \dot{\bigcup}_{j=1}^{c} (G \cap (t_j + \mathrm{Im}[\mathbb{I}])) .$$

PROOF. By Lemma 1.133, $\mathcal{N}_{\{o,o'\}}$ is a $\mathbb{Z}$-module of rank 6. Moreover, $\mathcal{N}_{\{o,o'\}}$ is obviously torsion-free. But $\mathrm{Im}[\mathbb{I}]$ is a $\mathbb{Z}$-module of rank 6 as well; see Corollary 1.100. Now, Proposition 1.4 yields the result. □

CHAPTER 2

Uniqueness

The problem of determining finite subsets of Delone sets $\varLambda$ by X-rays is considered. In particular, by using methods from convexity in conjunction with results from the previous chapter, sufficient conditions for the determination of the set of convex subsets of an algebraic Delone set $\varLambda$ by X-rays in four prescribed $\varLambda$-directions are provided. These results will also be applied to the special case of cyclotomic model sets, which also enables us to prove corresponding results in the case of icosahedral model sets. Further, a characterization of the affinely regular polygons in cyclotomic model sets in terms of a simple divisibility condition is provided. We also study the interactive technique of successive determination for cyclotomic model sets and icosahedral model sets, where the information from previous X-rays is used in deciding the direction for the next X-ray. In particular, it is shown that the finite subsets of these model sets $\varLambda$ can be successively determined by two $\varLambda$-directions.

2.1. General results on determination

In this section, we present some uniqueness results which are meant to motivate the later results of this chapter.

PROPOSITION 2.1. *Let $d \geq 2$ and let $\varLambda$ be a Delone set in $\mathbb{R}^d$. Further, let $u \in \mathbb{S}^{d-1}$ be a non-$\varLambda$-direction. Then, $\mathcal{F}(\varLambda)$ is determined by the single X-ray in direction u.*

PROOF. This follows immediately from the fact that a line in $\mathbb{R}^d$ in a non-$\varLambda$-direction passes through at most one point of $\varLambda$. □

REMARK 2.2. Since a Delone set $\varLambda \subset \mathbb{R}^d$ has only countably infinitely many elements, there are only countably infinitely many $\varLambda$-directions. The existence of non-$\varLambda$-direction follows. Note that Proposition 2.1 applies to model sets $\varLambda \subset \mathbb{R}^d$; see Remark 1.71.

The following result represents a fundamental source of difficulties in discrete tomography. There exist several versions; compare [**41**, Theorem 4.3.1] and [**32**, Lemma 2.3.2].

PROPOSITION 2.3. *Let $\varLambda \subset \mathbb{R}^2$ be an algebraic Delone set. Further, let $U \subset \mathbb{S}^1$ be an arbitrary, but fixed finite set of pairwise non-parallel $\varLambda$-directions. Then, $\mathcal{F}(\varLambda)$ is not determined by the X-rays in the directions of U.*

PROOF. By Lemma 1.62, we may assume, without loss of generality, that $0 \in \varLambda$. Consequently, we have $\mathbb{Q}(\varLambda) \subset \mathbb{Q}(\varLambda - \varLambda) \subset \mathbb{K}_{\varLambda}$. We argue by induction on $\mathrm{card}(U)$. The case $\mathrm{card}(U) = 0$ means $U = \varnothing$ and is obvious. Fix $k \in \mathbb{N}_0$ and suppose the assertion to be true whenever $\mathrm{card}(U) = k$. Let U now be a set with $\mathrm{card}(U) = k + 1$. By induction hypothesis, there are different elements F and F' of $\mathcal{F}(\varLambda)$ with the same X-rays in the directions of U', where $U' \subset U$ satisfies $\mathrm{card}(U') = k$. Let u be the remaining direction of U. Choose a non-zero element $z \in \mathbb{Q}(\varLambda - \varLambda)$ parallel to u such that $z + (F \cup F')$ and $F \cup F'$ are disjoint. Then,

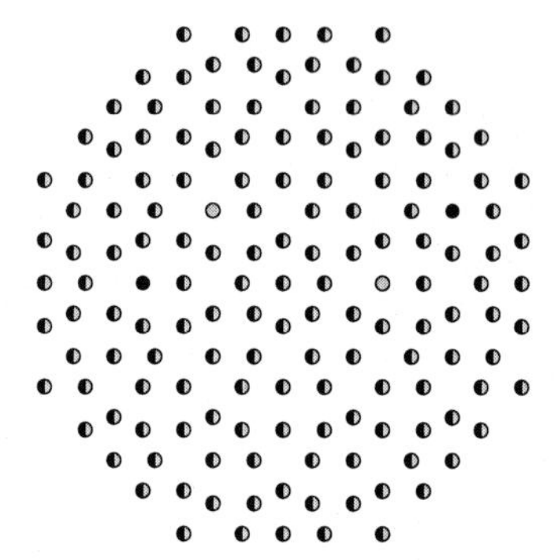

FIGURE 2.1. Two contiguous subsets of $\varLambda_{\mathrm{AB}}$ with the same X-rays in the $\mathcal{O}_8$-directions 1 and ζ_8.

$F'' := F \cup (z + F')$ and $F''' := F' \cup (z + F)$ are different elements of $\mathcal{F}(\mathbb{K}_\varLambda)$ with the same X-rays in the directions of U. By property (B) of algebraic Delone sets (see Definition 1.61), there is a homothety $h\colon \mathbb{R}^2 \longrightarrow \mathbb{R}^2$ such that $h[F'' \cup F'''] = h[F''] \cup h[F'''] \subset \varLambda$. It follows that $h[F'']$ and $h[F''']$ are different elements of $\mathcal{F}(\varLambda)$ with the same X-rays in the directions of U; see Lemma 1.113(b). □

REMARK 2.4. An analysis of the proof of Proposition 2.3 shows that, for any algebraic Delone set $\varLambda$ and for any finite set $U \subset \mathbb{S}^1$ of k pairwise non-parallel $\varLambda$-directions, there are disjoint elements F and F' of $\mathcal{F}(\varLambda)$ with $\mathrm{card}(F) = \mathrm{card}(F') = 2^{(k-1)}$ and with the same X-rays in the directions of U. Consider any convex (or bounded) set C in $\mathbb{R}^2$ which contains F and F' from above. Then, the subsets $F_1 := (C \cap \varLambda) \setminus F$ and $F_2 := (C \cap \varLambda) \setminus F'$ of $\mathcal{F}(\varLambda)$ also have the same X-rays in the directions of U. Whereas the points in F and F' are widely dispersed over a region, those in F_1 and F_2 are contiguous in a way similar to atoms in some condensed matter. This procedure is illustrated in Figure 2.1 in the case of the aperiodic cyclotomic model set $\varLambda_{\mathrm{AB}}$ associated with the Amman-Beenker tiling as described in Example 1.76; compare [**34**, Remark 4.3.2]. Note that, by Lemma 1.83, $\varLambda_{\mathrm{AB}}$ is an algebraic Delone set. Further, one can use Lemma 1.108 and Lemma 1.109(a) in order to show that the above argument extends to icosahedral model sets $\varLambda_{\mathrm{ico}}(t, W) \subset \mathbb{R}^3$ and finite sets $U \subset \mathbb{S}^2$ of pairwise non-parallel $\mathrm{Im}[\mathbb{I}]$-directions.

The proof of the following result can be the same as that of [**41**, Theorem 4.3.3]. Originally, the proof is due to Rényi; see [**56**]. For clarity, we prefer to repeat the details here, in a slightly modified way.

PROPOSITION 2.5. *Let $d \geq 2$ and let $\varLambda$ be a Delone set in $\mathbb{R}^d$. Further, let $U \subset \mathbb{S}^{d-1}$ be any set of $k+1$ pairwise non-parallel $\varLambda$-directions where $k \in \mathbb{N}_0$. Then, $\mathcal{F}_{\leq k}(\varLambda)$ is determined by the X-rays in the directions of U. Moreover, for all $F \in \mathcal{F}_{\leq k}(\varLambda)$, one has $G_U^F = F$.*

PROOF. Let $F, F' \in \mathcal{F}_{\leq k}(\Lambda)$ have the same X-rays in the directions of U. Then, one has

$$\operatorname{card}(F) = \operatorname{card}(F')$$

by Lemma 1.112(a) and

$$F, F' \subset G_U^F$$

by Lemma 1.120. But we have $G_U^F = F$ since the existence of a point in $G_U^F \setminus F$ implies the existence of at least $\operatorname{card}(U) \geq k+1$ points in F, a contradiction. It follows that $F = F'$. □

REMARK 2.6. Let Λ be an algebraic Delone set (e.g., a cyclotomic model set) or an icosahedral model set. Remark 2.4 and Proposition 2.5 show that $\mathcal{F}_{\leq k}(\Lambda)$ can be determined by the X-rays in any set of $k+1$ pairwise non-parallel Λ-directions but not by $1 + \lfloor \log_2 k \rfloor$ pairwise non-parallel X-rays in Λ-directions.

If, for the moment, one extends the concepts of determination and the definition of grids to the case of infinite sets of directions, an analysis of the proof of Proposition 2.5 gives the following result.

PROPOSITION 2.7. *Let $d \geq 2$ and let Λ be a Delone set in $\mathbb{R}^d$. Then, $\mathcal{F}(\Lambda)$ is determined by the set of (all) Λ-directions. Moreover, for all $F \in \mathcal{F}(\Lambda)$, one has $G_U^F = F$.* □

2.2. Determination of bounded subsets of Delone sets

2.2.1. Delone sets of finite local complexity.

THEOREM 2.8. *Let $d \geq 2$, let $R > 0$, and let $\Lambda \subset \mathbb{R}^d$ be a Delone set of finite local complexity. Then, one has:*

(a) *The set $\mathcal{D}_{<R}(\Lambda)$ is determined by two X-rays in Λ-directions.*
(b) *The set $\mathcal{D}_{<R}(\Lambda)$ is determined by two projections on orthogonal complements of 1-dimensional Λ-subspaces.*

PROOF. Let us first prove part (a). Since Λ has finite local complexity, there are only finitely many Λ-directions having the property that there is a set $F \in \mathcal{D}_{<R}(\Lambda)$ and a line ℓ in $\mathbb{R}^d$ in this direction with more than one point of F on ℓ. We denote the finite set of all these Λ-directions by U. Let $u \in \mathbb{S}^{d-1}$ be an arbitrary Λ-direction. For every $F \in \mathcal{D}_{<R}(\Lambda)$, Lemma 1.120 shows that

$$F \subset G_{\{u\}}^F \cap \Lambda .$$

Choose $u'' \in \mathbb{S}^{d-1} \cap (\mathbb{R}u)^{\perp}$ and note that, for every $F \in \mathcal{D}_{<R}(\Lambda)$, the set $(G_{\{u\}}^F \cap \Lambda)|(\mathbb{R}u)^{\perp}$ is finite with diameter

$$D_u^F := \operatorname{diam}\left((G_{\{u\}}^F \cap \Lambda)|(\mathbb{R}u)^{\perp}\right) < R .$$

Since Λ has finite local complexity, the set of diameters $\{D_u^F \mid F \in \mathcal{D}_{<R}(\Lambda)\}$ is finite. Set

$$D := \max\left(\{D_u^F \mid F \in \mathcal{D}_{<R}(\Lambda)\}\right) < R .$$

Note that there is an $\varepsilon_0 \in \mathbb{R}$ with $0 < \varepsilon_0 < 1$ such that every element of the set $B_{\varepsilon_0}(u'') \cap \mathbb{S}^{d-1}$ is a direction having the property that on each line in this direction there are no two points of any set $G_{\{u\}}^F \cap \Lambda$, $F \in \mathcal{D}_{<R}(\Lambda)$, on that line with a distance $\geq R$. Since the set of Λ-directions is dense in $\mathbb{S}^{d-1}$ by Lemma 1.59 (by assumption, Λ is a Delone set and hence relatively dense), and by the finiteness of the set U, this observation shows that one can choose a Λ-direction

non-parallel to u, say u', such that $u' \notin U$, and with the property that on each line in this direction there are no two points of any set $G^F_{\{u\}} \cap \Lambda$, $F \in \mathcal{D}_{<R}(\Lambda)$, on that line with a distance $\geq R$. We claim that $\mathcal{D}_{<R}(\Lambda)$ is determined by the X-rays in the directions u and u'. To see this, let $F, F' \in \mathcal{D}_{<R}(\Lambda)$ satisfy $X_u F = X_u F'$. Then, by Lemma 1.120, one has $F, F' \subset G^F_{\{u\}} \cap \Lambda$. In order to show that the identity $X_{u'} F = X_{u'} F'$ implies the equality $F = F'$, we shall even prove that each line in direction u' meets at most one point of $G^F_{\{u\}} \cap \Lambda$. Assume the existence of a line $\ell_{u'}$ in direction u', and assume the existence of two distinct points g and g' in $\ell_{u'} \cap (G^F_{\{u\}} \cap \Lambda)$. By construction, the distance of g and g' is less than R. Hence, one has $\{g, g'\} \in \mathcal{D}_{<R}(\Lambda)$, and further $u' \in U$, a contradiction. Part (b) follows immediately from an analysis of the proof of part (a). □

2.2.2. Application to Meyer sets and model sets.

Let us note some implications of Theorem 2.8.

COROLLARY 2.9. *Let $d \geq 2$, let $R > 0$, and let $\Lambda \subset \mathbb{R}^d$ be a Meyer set. Then, one has:*

(a) *The set $\mathcal{D}_{<R}(\Lambda)$ is determined by two X-rays in Λ-directions.*
(b) *The set $\mathcal{D}_{<R}(\Lambda)$ is determined by two projections on orthogonal complements of* 1-*dimensional Λ-subspaces.*

PROOF. This follows immediately from Theorem 2.8, since Meyer sets Λ are, by definition, Delone sets having the property that $\Lambda - \Lambda$ is uniformly discrete. Clearly, the latter property implies the finiteness of local complexity. □

COROLLARY 2.10. *Let $d \geq 2$, let $R > 0$, and let $\Lambda \subset \mathbb{R}^d$ be an arbitrary model set. Then, one has:*

(a) *The set $\mathcal{D}_{<R}(\Lambda)$ is determined by two X-rays in Λ-directions.*
(b) *The set $\mathcal{D}_{<R}(\Lambda)$ is determined by two projections on orthogonal complements of* 1-*dimensional Λ-subspaces.*

PROOF. This follows immediately from Corollary 2.9, since every model set is a Meyer set; compare [**48**]. □

COROLLARY 2.11. *Let $R > 0$ and let $\Lambda_{\text{ico}}(t, W) \subset \mathbb{R}^3$ be an icosahedral model set. Then, one has:*

(a) *The set $\mathcal{D}_{<R}(\Lambda_{\text{ico}}(t, W))$ is determined by two X-rays in* $\mathrm{Im}[\mathbb{I}]$-*directions.*
(b) *The set $\mathcal{D}_{<R}(\Lambda_{\text{ico}}(t, W))$ is determined by two projections on orthogonal complements of* 1-*dimensional* $\mathrm{Im}[\mathbb{I}]$-*subspaces.*

PROOF. This follows from Corollary 2.10 in conjunction with Lemma 1.109. □

COROLLARY 2.12. *Let $n \in \mathbb{N} \setminus \{1, 2\}$, let $\Lambda_n(t, W) \in \mathcal{M}(\mathcal{O}_n)$ be a cyclotomic model set, and let $R > 0$. Then, one has:*

(a) *The set $\mathcal{D}_{<R}(\Lambda_n(t, W))$ is determined by two X-rays in $\mathcal{O}_n$-directions.*
(b) *The set $\mathcal{D}_{<R}(\Lambda_n(t, W))$ is determined by two projections on orthogonal complements of* 1-*dimensional $\mathcal{O}_n$-subspaces.*

PROOF. This follows from Corollary 2.10 in conjunction with Lemma 1.84. □

REMARK 2.13. In Section 2.4, we shall prove a result on the *successive* determination of the whole set $\mathcal{F}(\Lambda_n(t,W))$ of finite subsets of a fixed cyclotomic model set $\Lambda_n(t,W)$ by X-rays or projections. There, we shall also give, independently of the investigations in this section, an alternative proof of Corollary 2.12.

COROLLARY 2.14. *Let $d \geq 2$, let $t \in \mathbb{R}^d$, and let $L \subset \mathbb{R}^d$ be a (full) lattice. Furthermore, let $R > 0$. Then, one has:*

(a) *The set $\mathcal{D}_{<R}(t+L)$ is determined by two X-rays in L-directions.*
(b) *The set $\mathcal{D}_{<R}(t+L)$ is determined by two projections on orthogonal complements of 1-dimensional L-subspaces.*

PROOF. This follows immediately from Corollary 2.10, since translates of lattices are model sets. □

REMARK 2.15. Clearly, Theorem 2.8 and the subsequent corollaries in this section are best possible with respect to the number of X-rays (resp., projections) used.

2.3. Determination of convex subsets of Delone sets by X-rays

2.3.1. Cross ratios.

DEFINITION 2.16. Let (t_1,t_2,t_3,t_4) be an ordered tuple of four distinct elements of the set $\mathbb{R}\cup\{\infty\}$. Then, its *cross ratio* $(t_1,t_2;t_3,t_4)$ is defined by

$$(t_1,t_2;t_3,t_4) := \frac{(t_3-t_1)(t_4-t_2)}{(t_3-t_2)(t_4-t_1)},$$

with the usual conventions if one of the t_i equals ∞, so $(t_1,t_2;t_3,t_4) \in \mathbb{R}$.

The following standard result is usually stated in the framework of projective geometry; compare [**19**, Ch. XI, in particular, Corollary 96.11]. For convenience, we give a reformulation and also include a proof.

LEMMA 2.17. *Let $z_j \in \mathbb{R}^2 \setminus \{0\}$, $j \in \{1,\dots,4\}$, be four pairwise non-parallel elements of the Euclidean plane with slopes $s_{z_j} \in \mathbb{R}\cup\{\infty\}$. Furthermore, let Ψ be a non-singular linear transformation of the plane. Then, one has*

$$(s_{z_1},s_{z_2};s_{z_3},s_{z_4}) = (s_{\Psi(z_1)},s_{\Psi(z_2)};s_{\Psi(z_3)},s_{\Psi(z_4)}).$$

PROOF. Let $z_j = (x_j,y_j)^t$, $j \in \{1,\dots,4\}$. Then, one has

$$(s_{z_1},s_{z_2};s_{z_3},s_{z_4}) = \frac{(\frac{y_3}{x_3}-\frac{y_1}{x_1})(\frac{y_4}{x_4}-\frac{y_2}{x_2})}{(\frac{y_3}{x_3}-\frac{y_2}{x_2})(\frac{y_4}{x_4}-\frac{y_1}{x_1})} = \frac{\det\left(\begin{smallmatrix}x_1 & x_3\\ y_1 & y_3\end{smallmatrix}\right)\det\left(\begin{smallmatrix}x_2 & x_4\\ y_2 & y_4\end{smallmatrix}\right)}{\det\left(\begin{smallmatrix}x_2 & x_3\\ y_2 & y_3\end{smallmatrix}\right)\det\left(\begin{smallmatrix}x_1 & x_4\\ y_1 & y_4\end{smallmatrix}\right)}. \tag{2.1}$$

The map $\Psi\colon \mathbb{R}^2 \longrightarrow \mathbb{R}^2$ is given by $z \longmapsto Az$, where A is a real 2×2 matrix with non-zero determinant. The assertion follows immediately from Equation (2.1) in conjunction with the multiplication theorem for determinants. □

DEFINITION 2.18. Let $z \in \mathbb{R}^2 \setminus \{0\}$, say $z = (x_z,y_z)^t$, and let $u \in \mathbb{S}^1$ be a direction.

(a) We denote by s_z the slope of z, i.e., $s_z = y_z/x_z \in \mathbb{R}\cup\{\infty\}$.
(b) The *angle between u and the positive real axis* is understood to be the unique angle $\theta \in [0,\pi)$ having the property that a rotation of $1 \in \mathbb{C}$ by θ in counter-clockwise order is a direction parallel to u.

LEMMA 2.19. *Let $\Lambda \subset \mathbb{R}^2$ and let $u \in \mathbb{S}^1$ be a Λ-direction. Then, one has*

$$s_u \in \Big(\mathbb{Q}\big((\Lambda - \Lambda) \cup (\overline{\Lambda - \Lambda}) \cup \{i\}\big) \cap \mathbb{R}\Big) \cup \Big\{\infty\Big\}.$$

PROOF. Let $u \in \mathbb{S}^1$ be a Λ-direction, say parallel to $z \in \Lambda - \Lambda \setminus \{0\}$. Then, one has

$$s_u = s_z = \frac{\frac{z-\bar{z}}{2i}}{\frac{z+\bar{z}}{2}} = -i\,\frac{z-\bar{z}}{z+\bar{z}} \in \Big(\mathbb{Q}\big((\Lambda - \Lambda) \cup (\overline{\Lambda - \Lambda}) \cup \{i\}\big) \cap \mathbb{R}\Big) \cup \Big\{\infty\Big\}. \tag{2.2}$$

The assertion follows. □

Astonishingly, one has the following result.

LEMMA 2.20. *For a set $\Lambda \subset \mathbb{R}^2$, the cross ratio of slopes of four pairwise non-parallel Λ-directions is an element of the field $\Bbbk_\Lambda$.*

PROOF. One can easily see from Equation (2.2) that, in the cross ratio of slopes of four pairwise non-parallel Λ-directions, the appearing terms of the form $-i$ can be cancelled out (even if one of the slopes equals ∞), hence the cross ratio is an element of the field $\Bbbk_\Lambda$. □

2.3.2. U-polygons in algebraic Delone sets. The following result was proved using Darboux's theorem [**25**] on midpoint polygons; see [**36**] or [**32**, Ch. 1] and compare [**34**, Lemma 4.3.6].

PROPOSITION 2.21. *If $U \subset \mathbb{S}^1$ is a finite set of directions, there exists a U-polygon if and only if there is an affinely regular polygon such that each direction of U is parallel to one of its edges.*

PROOF. See [**33**, Proposition 4.2]. □

REMARK 2.22. Note that a U-polygon need not be affinely regular, even if it is a U-polygon in an algebraic Delone set; see Example 2.46 below.

The proof of the following result is a modified version of that of [**33**, Lemma 4.4]; compare Remark 2.44.

LEMMA 2.23. *Let Λ be an algebraic Delone set. If $U \subset \mathbb{S}^1$ is any set of up to 3 pairwise non-parallel Λ-directions, then there exists a U-polygon in Λ.*

PROOF. Without loss of generality, we may assume that $\mathrm{card}(U) = 3$. First, construct a triangle in $\mathbb{K}_\Lambda$ having sides parallel to the given directions of U. If two of the vertices are chosen in $\mathbb{K}_\Lambda$, then the third is automatically in $\mathbb{K}_\Lambda$. Now, fit six congruent versions of this triangle together in the obvious way to make an affinely regular hexagon in $\mathbb{K}_\Lambda$. The latter is then a U-polygon, say P, in $\mathbb{K}_\Lambda$. By property (B) of algebraic Delone sets, there is a homothety $h\colon \mathbb{R}^2 \longrightarrow \mathbb{R}^2$ such that $P' := h[P]$ is a polygon in Λ. Since P' is a U-polygon (see Lemma 1.113(a)), P' is a U-polygon in Λ. □

Recall Definition 1.37 on the sets D_m and the mappings f_m. Suppose the existence of a U-polygon in an algebraic Delone set Λ. Then, the set U consists of Λ-directions. The proof of the following result is a modified version of the first part of the proof of [**33**, Theorem 4.5].

THEOREM 2.24. *Let $\Lambda \subset \mathbb{R}^2$, let $U \subset \mathbb{S}^1$ be a set of four or more pairwise non-parallel Λ-directions, and suppose the existence of a U-polygon. Then, the cross ratio of slopes of any four directions of U, arranged in order of increasing angle with the positive real axis, is an element of the set*

$$\Big(\bigcup_{m\geq 4} f_m[D_m]\Big) \cap \mathbb{k}_\Lambda .$$

PROOF. Let U be as in the assertion. By Proposition 2.21, U consists of directions parallel to the edges of an affinely regular polygon. Hence, there is a non-singular linear transformation Ψ of the plane with the following property: If one sets

$$V := \big\{ u_{\Psi(u')} \,\big|\, u' \in U \big\} \subset \mathbb{S}^1 ,$$

then V is contained in a set of directions that are equally spaced in $\mathbb{S}^1$, i.e., the angle between each pair of adjacent directions is the same. Since the directions of U are pairwise non-parallel, there is an $m \in \mathbb{N}$ with $m \geq 4$ such that each direction of V is parallel to a direction of the form $e^{h\pi i/m}$, where $h \in \mathbb{N}_0$ satisfies $h \leq m-1$. Let u'_j, $1 \leq j \leq 4$, be four directions of U, arranged in order of increasing angle with the positive real axis. By Lemma 2.20, the cross ratio of the slopes of these Λ-directions, say $q := (s_{u'_1}, s_{u'_2}; s_{u'_3}, s_{u'_4})$, is an element of the real algebraic number field $\mathbb{k}_\Lambda$. One can see by Lemma 2.17 together with the fact that every non-singular linear transformation of the plane either preserves or inverts orientation that we may assume, without loss of generality, that each direction $u_{\Psi(u'_j)} \in V$ is parallel to a direction of the form $e^{h_j\pi i/m}$, where $h_j \in \mathbb{N}_0$, $1 \leq j \leq 4$, and, moreover, $h_1 < h_2 < h_3 < h_4 \leq m-1$. Using Lemma 2.17 again, one gets

$$q = (s_{\Psi(u'_1)}, s_{\Psi(u'_2)}; s_{\Psi(u'_3)}, s_{\Psi(u'_4)}) = \frac{(\tan(\frac{h_3\pi}{m}) - \tan(\frac{h_1\pi}{m}))(\tan(\frac{h_4\pi}{m}) - \tan(\frac{h_2\pi}{m}))}{(\tan(\frac{h_3\pi}{m}) - \tan(\frac{h_2\pi}{m}))(\tan(\frac{h_4\pi}{m}) - \tan(\frac{h_1\pi}{m}))} .$$

Manipulating the right-hand side, one obtains

$$q = \frac{\sin(\frac{(h_3-h_1)\pi}{m})\sin(\frac{(h_4-h_2)\pi}{m})}{\sin(\frac{(h_3-h_2)\pi}{m})\sin(\frac{(h_4-h_1)\pi}{m})} .$$

Setting $k_1 := h_3 - h_1$, $k_2 := h_4 - h_2$, $k_3 := h_3 - h_2$ and $k_4 := h_4 - h_1$, one gets $1 \leq k_3 < k_1, k_2 < k_4 \leq m-1$ and $k_1 + k_2 = k_3 + k_4$.

Using $\sin(\theta) = \frac{-e^{-i\theta}(1-e^{2i\theta})}{2i}$, one obtains

$$\mathbb{k}_\Lambda \ni q = \frac{(1-\zeta_m^{k_1})(1-\zeta_m^{k_2})}{(1-\zeta_m^{k_3})(1-\zeta_m^{k_4})} = f_m(d) ,$$

with $d := (k_1, k_2, k_3, k_4)$, as in (1.11). Then, $d \in D_m$ if its first two coordinates are interchanged, if necessary, to ensure that $k_1 \leq k_2$; note that this operation does not change the value of $f_m(d)$; see Definition 1.37. This completes the proof. □

We are now able to prove the following results.

THEOREM 2.25. *For any $e \in \mathbb{N}$, there is a finite set $N_e \subset \mathbb{Q}$ such that, for all algebraic Delone sets Λ with $[\mathbb{k}_\Lambda : \mathbb{Q}] = e$ and all sets $U \subset \mathbb{S}^1$ of four or more pairwise non-parallel Λ-directions, one has the following:*

If there exists a U-polygon, then the cross ratio of slopes of any four directions of U, arranged in order of increasing angle with the positive real axis, maps under the norm $N_{\mathbb{k}_\Lambda/\mathbb{Q}}$ to N_e.

PROOF. First, note that, by Lemma 1.64, every algebraic Delone sets Λ satisfies the relation $[\mathbb{k}_\Lambda : \mathbb{Q}] < \infty$. The assertion is an immediate consequence of Theorem 2.24 in conjunction with Theorem 1.49. □

THEOREM 2.26. *For all algebraic Delone sets Λ and all sets $U \subset \mathbb{S}^1$ of four or more pairwise non-parallel Λ-directions, one has the following:*

If there exists a U-polygon, then the cross ratio of slopes of any four directions of U, arranged in order of increasing angle with the positive real axis, maps under the norm $N_{\mathbb{k}_\Lambda/\mathbb{Q}}$ to the set

$$\pm\left(\{1\} \cup \left[\mathbb{Q}_{>1}^{\mathbb{P}_{\leq 2}}\right]^{\pm 1}\right).$$

PROOF. First, note that, by Lemma 1.64, every algebraic Delone sets Λ satisfies the relation $[\mathbb{k}_\Lambda : \mathbb{Q}] < \infty$. The assertion is an immediate consequence of Theorem 2.24 in conjunction with Theorem 1.51. □

2.3.3. Determination of convex subsets of algebraic Delone sets by X-rays.

LEMMA 2.27. *Let $\Lambda \subset \mathbb{R}^2$ be a Delone set and let $U \subset \mathbb{S}^1$ be a finite set of at least three pairwise non-parallel Λ-directions. Suppose the existence of $F, F' \in \mathcal{C}(\Lambda)$ such that $X_u F = X_u F'$ for all $u \in U$. Then, one has*

$$F \neq F' \implies \dim(F) = \dim(F') = 2.$$

PROOF. No changes needed in comparison with the proof of [**33**, Lemma 5.2]. □

REMARK 2.28. In general, Lemma 2.27 is false if one reduces the number of pairwise non-parallel Λ-directions of U to two.

The proof of the following result is a modified version of that of [**33**, Theorem 5.5].

THEOREM 2.29. *Let Λ be an algebraic Delone set and let $U \subset \mathbb{S}^1$ be a set of two or more pairwise non-parallel Λ-directions. The following statements are equivalent:*

(i) *$\mathcal{C}(\Lambda)$ is determined by the X-rays in the directions of U.*
(ii) *There is no U-polygon in Λ.*

PROOF. By Lemma 1.62, we may assume, without loss of generality, that $0 \in \Lambda$, whence $\Lambda \subset \Lambda - \Lambda$. For (i) $\Rightarrow$ (ii), suppose the existence of a U-polygon P in Λ. Partition the vertices of P into two disjoint sets V, V', where the elements of these sets alternate round the boundary $\mathrm{bd}(P)$ of P. Since P is a U-polygon, each line in the plane parallel to some $u \in U$ that contains a point in V also contains a point in V'. In particular, one sees that $\mathrm{card}(V) = \mathrm{card}(V')$. Set

$$C := (\Lambda \cap P) \setminus (V \cup V')$$

and further $F_1 := C \cup V$ and $F_2 := C \cup V'$. Then, F_1 and F_2 are different convex subsets of Λ with the same X-rays in the directions of U.

For (ii) $\Rightarrow$ (i), suppose that F_1 and F_2 are different convex subsets of $\varLambda$ with the same X-rays in the directions of U. Set

$$E := \operatorname{conv}(F_1) \cap \operatorname{conv}(F_2).$$

We may assume that $\operatorname{card}(U) \geq 4$, since Lemma 2.23 provides a U-polygon in $\varLambda$ whenever $\operatorname{card}(U) \leq 3$. By Lemma 2.27, we have $\dim(F_1) = \dim(F_2) = 2$ and Lemma 1.112(b) shows that F_1 and F_2 have the same centroid. It follows that $\operatorname{int}(E) \neq \varnothing$.

Since $\operatorname{conv}(F_1)$ and $\operatorname{conv}(F_2)$ are convex polygons, one knows that

$$\operatorname{int}\left(\operatorname{conv}(F_1) \bigtriangleup \operatorname{conv}(F_2)\right)$$

has finitely many components. By the assumption $F_1 \neq F_2$, there is at least one component. Let these components be C_j, and call C_j of type $r \in \{1,2\}$ if $C_j \subset \operatorname{int}(\operatorname{conv}(F_r) \setminus E)$. Consider the set of type 1 (resp., type 2) components together with the equivalence relation generated by the reflexive and symmetric relation R given by adjacency, i.e., $C\,R\,C' \iff \operatorname{cl}(C) \cap \operatorname{cl}(C') \neq \varnothing$. Let the set $\mathcal{D}_1$ (resp., $\mathcal{D}_2$) consist of all unions $\cup \mathcal{C}$, where $\mathcal{C}$ is an equivalence class of type 1 (resp., type 2) components. Let $\mathcal{D} := \mathcal{D}_1 \cup \mathcal{D}_2$. Note that the elements of $\mathcal{D}_1$ and $\mathcal{D}_2$ alternate round the boundary $\operatorname{bd}(E)$ of E.

Suppose that $D \in \mathcal{D}_1$. The set $A := (\operatorname{cl}(D) \setminus E) \cap \varLambda$ is non-empty, finite and contained in $F_1 \setminus E$. If $u \in U$ and $z \in A$, then, since $X_u F_1 = X_u F_2$, there is an element $z' \in \varLambda$ which satisfies

$$z' \in (F_2 \setminus E) \cap \ell_u^z .$$

It follows that ℓ_u^z meets some element of $\mathcal{D}_2$. Let us denote this element by $D(u)$.

We first claim that $D(u)$ does not depend on the choice of $z \in A$. To see this, let $\tilde{z} \in A$ be another element of A (i.e., $z \neq \tilde{z} \in A$) such that $\ell_u^{\tilde{z}}$ meets $\tilde{D}(u) \in \mathcal{D}_2$, where $\tilde{D}(u) \neq D(u)$. The latter inequality implies that $\tilde{D}(u)$ and $D(u)$ are disjoint and, moreover, we see that with respect to the clockwise ordering round $\operatorname{bd}(E)$ there exists an element D' of $\mathcal{D}_1$ between $\tilde{D}(u)$ and $D(u)$. There follows the existence of an element $\hat{z} \in \varLambda$ contained in the open strip bounded by ℓ_u^z and $\ell_u^{\tilde{z}}$ such that $\hat{z} \in \operatorname{cl}(C) \setminus E$, where C is one of the type 1 components contained in D'. Since $X_u F_1 = X_u F_2$, there follows the existence of an element $\hat{z}' \in \varLambda \cap \ell_u^{\hat{z}}$ with $\hat{z}' \in \operatorname{cl}(C') \setminus E$, where C' is a type 2 component. It follows that $C' \subset D$, a contradiction. This proves the claim.

The set $A(u) := (\operatorname{cl}(D(u)) \setminus E) \cap \varLambda$ is finite and contained in $F_2 \setminus E$. Moreover, since $X_u A(u) = X_u A$, we have $\operatorname{card}(A(u)) = \operatorname{card}(A)$ by Lemma 1.112(a). In particular, we see that $A(u)$ is non-empty.

By symmetry, one gets analogous results for any element $D \in \mathcal{D}_2$. Choose an arbitrary $D \in \mathcal{D}$ and define the subset

$$\mathcal{D}' := \left\{ ((\dots(D(u'_{i_1}))\dots)(u'_{i_{k-1}}))(u'_{i_k}) \,|\, k \in \mathbb{N}, u'_{i_j} \in U \text{ for all } j \in \{1,\dots,k\} \right\}.$$

of $\mathcal{D}$, obtained from D by applying the above process through any finite sequence of directions from U. Let $\mathcal{D}' = \{D_j \,|\, j \in \{1,\dots,m\}\}$ and let $A_j := (\operatorname{cl}(D_j) \setminus E) \cap \varLambda$ be the non-empty set of elements of $\varLambda$ corresponding to D_j, $j \in \{1,\dots,m\}$.

Let c_j be the centroid of A_j, $j \in \{1,\dots,m\}$, and let t_j be the line through the common endpoints of the two arcs, one in $\operatorname{bd}(\operatorname{conv}(F_1))$, the other in $\operatorname{bd}(\operatorname{conv}(F_2))$, which bound D_j. Then, t_j separates A_j, and hence c_j, from the convex hull of the remaining centroids c_k, with $k \in \{1,\dots,m\} \setminus \{j\}$. It follows that the points c_j, $j \in \{1,\dots,m\}$, are the vertices of a convex polygon P. If $u \in U$ and $j \in \{1, \ldots, m\}$, suppose that A_k is the set arising from u and A_j by the process described above, i.e., $A_k = A_j(u)$. Then, by Lemma 1.112(b), the line $\ell_u^{c_j}$ also contains c_k. The points c_j therefore pair off in this fashion, so m is even, and since $\mathrm{card}(U) \geq 2$, we have $m \geq 4$, and P is non-degenerate. Hence, P is a U-polygon.

Let $\mathrm{card}(A_1) = \cdots = \mathrm{card}(A_m) =: s \in \mathbb{N}$. Then, each vertex of P belongs to the set

$$\Big\{\frac{1}{s}(\sum_{j=1}^{s} \lambda_j) \,\Big|\, \lambda_1, \ldots, \lambda_s \in \Lambda\Big\} \subset \mathbb{Q}(\Lambda) \subset \mathbb{Q}(\Lambda - \Lambda) \subset \mathbb{K}_\Lambda .$$

Hence, P is a U-polygon in $\mathbb{K}_\Lambda$. By property (B) of algebraic Delone sets, there is a homothety $h : \mathbb{R}^2 \longrightarrow \mathbb{R}^2$ such that $P' := h[P]$ is a polygon in Λ. Since P' is a U-polygon (see Lemma 1.113(a)), P' is a U-polygon in Λ. □

REMARK 2.30. In the proof of Theorem 2.29, it is necessary to employ finite unions of components; compare [**33**, Remark 5.6].

We are now able to prove the following central results.

THEOREM 2.31. *The following statements hold:*

(a) *For any $e \in \mathbb{N}$, there is a finite set $N_e \subset \mathbb{Q}$ such that, for all algebraic Delone sets Λ with $[\mathbb{k}_\Lambda : \mathbb{Q}] = e$ and for all sets $U \subset \mathbb{S}^1$ of four pairwise non-parallel Λ-directions, one has the following:*

  If U has the property that the cross ratio of slopes of the directions of U, arranged in order of increasing angle with the positive real axis, does not map under the norm $N_{\mathbb{k}_\Lambda/\mathbb{Q}}$ to N_e, then $C(\Lambda)$ is determined by the X-rays in the directions of U.

(b) *For all algebraic Delone sets Λ and all sets $U \subset \mathbb{S}^1$ of three or less pairwise non-parallel Λ-directions, the set $C(\Lambda)$ is not determined by the X-rays in the directions of U.*

PROOF. Part (a) follows immediately from Theorem 2.29 in conjunction with Theorem 2.25. Assertion (b) is an immediate consequence of Theorem 2.29 in conjunction with Lemma 2.23. □

Applying Theorem 2.29 in conjunction with Theorem 2.26 gives:

THEOREM 2.32. *For all algebraic Delone set Λ and all sets $U \subset \mathbb{S}^1$ of four pairwise non-parallel Λ-directions, one has the following:*

If U has the property that the cross ratio of slopes of the directions of U, arranged in order of increasing angle with the positive real axis, does not map under the norm $N_{\mathbb{k}_\Lambda/\mathbb{Q}}$ to the set

$$\pm\left(\{1\} \cup \left[\mathbb{Q}_{>1}^{\mathbb{P}_{\leq 2}}\right]^{\pm 1}\right),$$

then $C(\Lambda)$ is determined by the X-rays in the directions of U. □

2.3.4. Application to cyclotomic model sets and icosahedral model sets.

2.3.4.1. *Affinely regular polygons in cyclotomic model sets.* R. J. Gardner and P. Gritzmann [**33**, Theorem 4.1] have shown that there is an affinely regular m-gon in the square lattice $\mathbb{Z}^2 = \mathbb{Z}[\zeta_4] = \mathcal{O}_4$ if and only if $m \in \{3, 4, 6\}$. We start off with a generalization of this result to rings of cyclotomic integers.

THEOREM 2.33. *Let $m, n \in \mathbb{N}$ with $m, n \geq 3$. The following statements are equivalent:*

(i) *There is an affinely regular m-gon in $\mathcal{O}_n$.*
(ii) $\Bbbk_m \subset \Bbbk_n$.
(iii) $m \in \{3,4,6\}$, *or* $\mathbb{K}_m \subset \mathbb{K}_n$.
(iv) $m \in \{3,4,6\}$, *or* $m|n$, *or* $m = 2d$ *with* d *an odd divisor of* n.
(v) $m \in \{3,4,6\}$, *or* $\mathcal{O}_m \subset \mathcal{O}_n$.
(vi) $\mathcal{o}_m \subset \mathcal{o}_n$.

PROOF. For (i) $\Rightarrow$ (ii), let P be an affinely regular m-gon in $\mathcal{O}_n$. There is then a non-singular affine transformation $\Psi : \mathbb{R}^2 \longrightarrow \mathbb{R}^2$ with $\Psi[R_m] = P$, where R_m is the regular m-gon with vertices given in complex form by $1, \zeta_m, \dots, \zeta_m^{m-1}$. If $m \in \{3,4,6\}$, condition (ii) holds trivially. Suppose $6 \neq m \geq 5$. The pairs $\{1, \zeta_m\}$, $\{\zeta_m^{-1}, \zeta_m^2\}$ lie on parallel lines and so do their images under Ψ. Therefore,

$$\frac{\|\zeta_m^2 - \zeta_m^{-1}\|}{\|\zeta_m - 1\|} = \frac{\|\Psi(\zeta_m^2) - \Psi(\zeta_m^{-1})\|}{\|\Psi(\zeta_m) - \Psi(1)\|}.$$

Moreover, we get the relation

$$(1 + \zeta_m + \bar{\zeta}_m)^2 = (1 + \zeta_m + \zeta_m^{-1})^2 = \frac{\|\zeta_m^2 - \zeta_m^{-1}\|^2}{\|\zeta_m - 1\|^2} = \frac{\|\Psi(\zeta_m^2) - \Psi(\zeta_m^{-1})\|^2}{\|\Psi(\zeta_m) - \Psi(1)\|^2} \in \Bbbk_n \,.$$

The pairs $\{\zeta_m^{-1}, \zeta_m\}$, $\{\zeta_m^{-2}, \zeta_m^2\}$ also lie on parallel lines. An argument similar to that above yields

$$(\zeta_m + \bar{\zeta}_m)^2 = (\zeta_m + \zeta_m^{-1})^2 = \frac{\|\zeta_m^2 - \zeta_m^{-2}\|^2}{\|\zeta_m - \zeta_m^{-1}\|^2} \in \Bbbk_n \,.$$

By subtracting these equations, we get

$$2(\zeta_m + \bar{\zeta}_m) + 1 \in \Bbbk_n \,,$$

and hence $\zeta_m + \bar{\zeta}_m \in \Bbbk_n$, the latter being equivalent to the inclusion of the fields $\Bbbk_m \subset \Bbbk_n$.

As an immediate consequence of Lemma 1.20(b), we get (ii) $\Rightarrow$ (iii). Conditions (iii) and (iv) are equivalent by Lemma 1.17, and (iii) $\Rightarrow$ (v) $\Rightarrow$ (vi) follows from Proposition 1.23.

For (vi) $\Rightarrow$ (i), let R_m again be the regular m-gon as defined in the step (i) $\Rightarrow$ (ii). Since $m, n \geq 3$, the sets $\{1, \zeta_m\}$ and $\{1, \zeta_n\}$ are $\mathbb{R}$-bases of $\mathbb{C}$. Hence, we can define an $\mathbb{R}$-linear map $L_m^n : \mathbb{R}^2 \longrightarrow \mathbb{R}^2$ as the linear extension of $1 \longmapsto 1$ and $\zeta_m \longmapsto \zeta_n$. Obviously, L_m^n is non-singular. Then, using $\mathcal{o}_m \subset \mathcal{o}_n$, one can see, by means of Lemma 1.9(a), that the vertices of $L_m^n[R_m]$, i.e., $L_m^n(1), L_m^n(\zeta_m), \dots, L_m^n(\zeta_m^{m-1})$, lie in $\mathcal{O}_n$ (in fact, L_m^n maps the whole $\mathcal{o}_m$-module $\mathcal{O}_m$ into the $\mathcal{o}_n$-module $\mathcal{O}_n$), whence $L_m^n[R_m]$ is a polygon in $\mathcal{O}_n$. This shows that there is an affinely regular m-gon in $\mathcal{O}_n$. □

REMARK 2.34. Alternatively, there is the following direct argument for (v) $\Rightarrow$ (i) in Theorem 2.33. If $m \in \{3,4,6\}$, then $L_m^n[R_m]$ from the proof of (vi) $\Rightarrow$ (i) above is an affinely regular m-gon in $\mathcal{O}_n$. Otherwise, one can simply choose the regular m-gon R_m (formed by the mth roots of unity) itself.

Although the equivalence (i) $\Leftrightarrow$ (iv) in Theorem 2.33 is a satisfactory characterization, the following corollary deals with the two cases where condition (ii) can be used more effectively.

COROLLARY 2.35. *Let $m \in \mathbb{N}$ with $m \geq 3$. Consider ϕ on $\{n \in \mathbb{N} \,|\, n \not\equiv 2 \pmod 4\}$. Then, one has:*

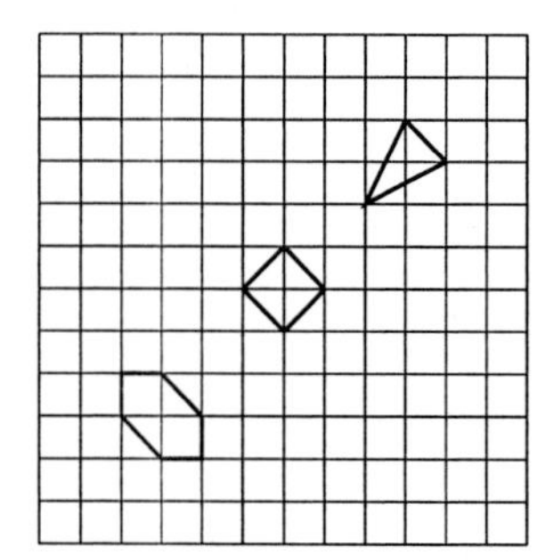

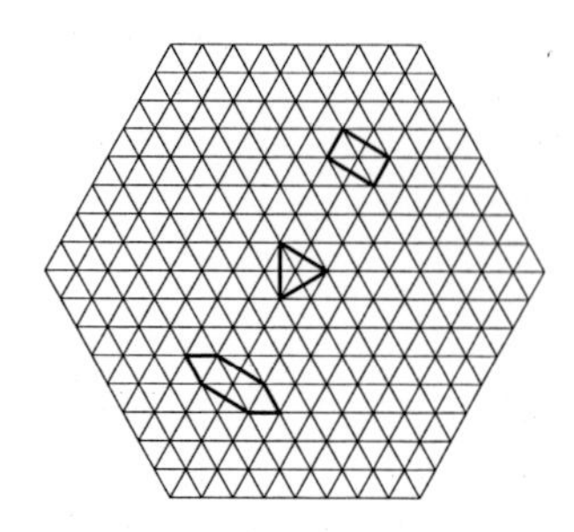

FIGURE 2.2. On the left, a central patch of the square tiling with vertex set Λ_{SQ} with $L_4^4[R_4]$ and translates of $L_3^4[R_3]$ and $L_6^4[R_6]$ as described in example (SQ) is shown. On the right, a central patch of the triangular tiling with vertex set Λ_{TRI} with $L_3^3[R_3]$ and translates of $L_4^3[R_4]$ and $L_6^3[R_6]$ as described in example (TRI) is shown.

(a) *If $\phi(n)/2 = 1$ (i.e., $n \in \{3,4\}$; see Lemma 1.21(a)), there is an affinely regular m-gon in $\mathcal{O}_n$ if and only if $m \in \{3,4,6\}$, i.e., the affinely regular polygons in $\mathcal{O}_n$ in this case are exactly the affinely regular triangles, parallelograms and hexagons.*

(b) *If $\phi(n)/2 \in \mathbb{P}$ (i.e., $n \in \mathbb{S}$; see Lemma 1.21(b)), there is an affinely regular m-gon in $\mathcal{O}_n$ if and only if*

$$\begin{cases} m \in \{3,4,6,n\}, & \text{if } n = 8 \text{ or } n = 12, \\ m \in \{3,4,6,n,2n\}, & \text{otherwise.} \end{cases}$$

PROOF. If $\phi(n)/2 = 1$, condition (ii) of Theorem 2.33 specializes to $\Bbbk_m = \mathbb{Q}$. This is equivalent to $\phi(m) = 2$ by Corollary 1.14, which means $m \in \{3,4,6\}$. This proves the part (a).

If $\phi(n)/2 \in \mathbb{P}$, we have $[\Bbbk_n : \mathbb{Q}] = \phi(n)/2 \in \mathbb{P}$. Hence, we get, by means of condition (ii) of Theorem 2.33 and Lemma 1.13, either $\Bbbk_m = \mathbb{Q}$ or $\Bbbk_m = \Bbbk_n$. The former case implies $m \in \{3,4,6\}$ as in the proof of the part (a), while the proof follows from Lemma 1.20(a) in conjunction with Corollary 1.18 in the latter case. □

COROLLARY 2.36. *Let $m \in \mathbb{N}$ with $m \geq 3$. Further, let $n \in \mathbb{N} \setminus \{1,2\}$ and let $\Lambda_n(t,W) \in \mathcal{M}(\mathcal{O}_n)$ be a cyclotomic model set. The following statements are equivalent:*

(i) *There is an affinely regular m-gon in $\Lambda_n(t,W)$.*
(ii) $\Bbbk_m \subset \Bbbk_n$.
(iii) *$m \in \{3,4,6\}$, or $\mathbb{K}_m \subset \mathbb{K}_n$.*
(iv) *$m \in \{3,4,6\}$, or $m|n$, or $m = 2d$ with d an odd divisor of n.*
(v) *$m \in \{3,4,6\}$, or $\mathcal{O}_m \subset \mathcal{O}_n$.*
(vi) $\mathcal{O}_m \subset \mathcal{O}_n$.

PROOF. The assertion is a consequence of Theorem 2.33 and Lemma 1.81, since homotheties are non-singular affine transformations. □

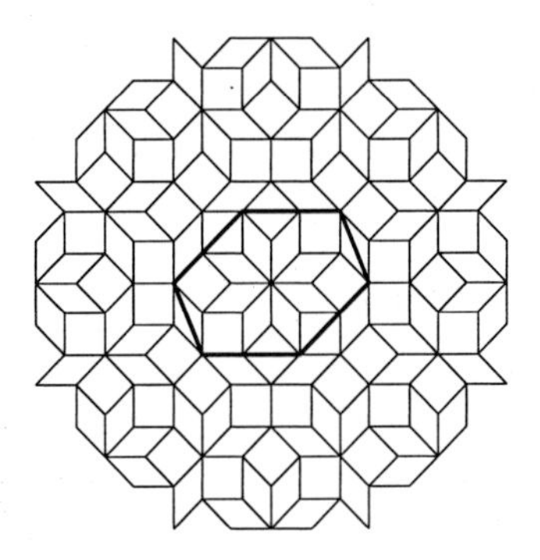

FIGURE 2.3. A central patch of the eightfold symmetric Ammann-Beenker tiling with vertex set Λ_{AB}. In the tiling, the affinely regular 6-gon as described in example (AB) is shown, the other solutions being rather obvious.

COROLLARY 2.37. *Let $m \in \mathbb{N}$ with $m \geq 3$. Further, let $n \in \mathbb{S}$ and let $\Lambda_n(t,W) \in \mathcal{M}(\mathcal{O}_n)$ be a (aperiodic) cyclotomic model set. Then, there is an affinely regular m-gon in $\Lambda_n(t,W)$ if and only if*

$$\begin{cases} m \in \{3,4,6,n\}, & \text{if } n = 8 \text{ or } n = 12, \\ m \in \{3,4,6,n,2n\}, & \end{cases}$$

PROOF. This follows immediately from Corollary 2.35(b) and Lemma 1.81, since homotheties are non-singular affine transformations.. □

EXAMPLE 2.38. In the following, we illustrate our results above for the cyclotomic model sets Λ_{SQ}, Λ_{TRI}, Λ_{AB}, Λ_{TTT} and Λ_{S} (as defined in Example 1.76). Moreover, we assume that R_m and $L_m^n[R_m]$ are as defined in Remark 2.34.

(SQ) By Corollary 2.35(a), there is an affinely regular m-gon in $\mathcal{O}_4$ if and only if $m \in \{3,4,6\}$. Moreover, Remark 2.34 shows that the affinely regular 3-, 4- and 6-gons $L_3^4[R_3]$, $L_4^4[R_4]$ and $L_6^4[R_6]$ are polygons in $\mathcal{O}_4$; see Figure 2.2.

(TRI) Again by Corollary 2.35(a), there is an affinely regular m-gon in $\mathcal{O}_3$ if and only if $m \in \{3,4,6\}$. Moreover, Remark 2.34 shows that the affinely regular 3-, 4- and 6-gons $L_3^3[R_3]$, $L_4^3[R_4]$ and $L_6^3[R_6]$ are polygons in $\mathcal{O}_3$; see Figure 2.2.

All further examples below are aperiodic cyclotomic model sets of the form $\Lambda_n(0,W) \in \mathcal{M}(\mathcal{O}_n)$ for suitable $n \in \mathbb{S}$ and satisfy $0 \in \mathrm{int}(W)$. An analysis of the proof of Lemma 1.81 in connection with Corollary 2.37 and Remark 2.34 shows the existence of an expansive dilatation $d\colon \mathbb{R}^2 \longrightarrow \mathbb{R}^2$, given by multiplication by a suitable non-negative integral power of a PV-number of $\mathcal{O}_n$, such that $d[L_m^n[R_m]] \subset \Lambda_n(0,W)$, if $m \in \{3,4,6\}$ (respectively, $d[R_m] \subset \Lambda_n(0,W)$, otherwise), with n, m in accordance with the statement of Corollary 2.37. Let us demonstrate this in some more detail.

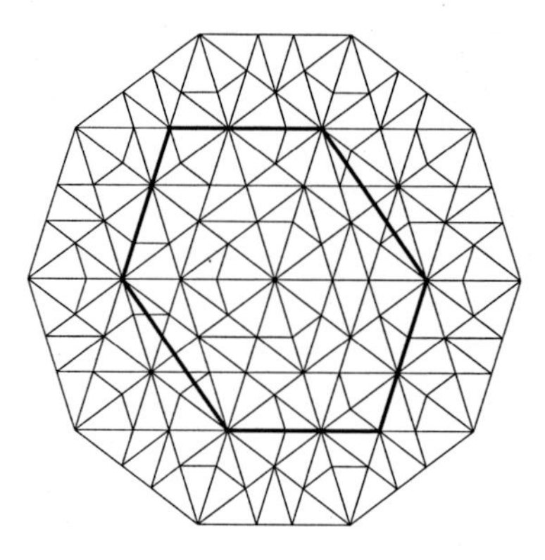

FIGURE 2.4. A central patch of the tenfold symmetric Tübingen triangle tiling. Therein, the affinely regular 6-gon as described in example (TTT) is marked, the other solutions being rather obvious.

(AB) By Corollary 2.37, there is an affinely regular m-gon in $\varLambda_{\mathrm{AB}}$ if and only if $m \in \{3,4,6,8\}$. See Figure 2.3 for all assertions below. The affinely regular 3-, 4- and 8-gons $L_3^8[R_3]$, $L_4^8[R_4]$ and R_8 are polygons in $\varLambda_{\mathrm{AB}}$. For an affinely regular 6-gon with its vertices in $\varLambda_{\mathrm{AB}}$, consider the expansive dilatation of $L_6^8[R_6]$, which is given by multiplication by the PV-unit $1+\sqrt{2}$ in $\mathcal{O}_8$ (the fundamental unit in $\mathcal{O}_8$), i.e., the convex 6-gon with vertices $1+\sqrt{2}$, $(1+\sqrt{2})\zeta_8$, $(1+\sqrt{2})(-1+\zeta_8)$, $-(1+\sqrt{2})$, $-(1+\sqrt{2})\zeta_8$, $(1+\sqrt{2})(1-\zeta_8)$. Here, also the PV-number $2+\sqrt{2}$ of (full) degree 2 would be suitable.

(TTT) Again by Corollary 2.37, there is an affinely regular m-gon in $\varLambda_{\mathrm{TTT}}$ if and only if $m \in \{3,4,5,6,10\}$. See Figure 2.4 for all assertions below. The affinely regular 4-, 5- and 10-gons $L_4^5[R_4]$, R_5 and R_{10} are polygons in $\varLambda_{\mathrm{TTT}}$. For an affinely regular polygon in $\varLambda_{\mathrm{TTT}}$ with 3 vertices, consider the expansive dilatation of $L_3^5[R_3]$, which is given by multiplication by the PV-unit τ^2 in $\mathcal{O}_5$, i.e., the convex 3-gon with vertices $\tau^2, \tau^2\zeta_5, \tau^2(-1-\zeta_5)$. Note the identity $\tau^2 = 2+\frac{1}{\tau}$. Note further that also the PV-unit $1+\frac{1}{\tau} = \tau$ would be suitable. For an affinely regular polygon in $\varLambda_{\mathrm{TTT}}$ with 6 vertices, consider the expansive dilatation of $L_6^5[R_6]$, which is again given by multiplication with the PV-unit $\tau^2 = 2+1/\tau$ in $\mathcal{O}_5$, i.e., the convex 6-gon with vertices τ^2, $\tau^2\zeta_5$, $\tau^2(-1+\zeta_5)$, $-\tau^2$, $-\tau^2\zeta_5$, $\tau^2(1-\zeta_5)$. Again, also the PV-unit $1+1/\tau = \tau$ would be suitable.

(S) Once more, by Corollary 2.37, there is an affinely regular m-gon in $\varLambda_{\mathrm{S}}$ if and only if $m \in \{3,4,6,12\}$. See Figure 2.5 for all assertions below. It is immediate that $L_3^{12}[R_3]$, $L_4^{12}[R_4]$ and R_{12} are polygons in $\varLambda_{\mathrm{S}}$. For an affinely regular polygon in $\varLambda_{\mathrm{S}}$ with 6 vertices, consider the expansive dilatation of $L_6^{12}[R_6]$ which is given by multiplication with the PV-number $1+\sqrt{3}$ of (full) degree 2 in $\mathcal{O}_{12}$, i.e., the convex 6-gon with vertices $1+\sqrt{3}$, $(1+\sqrt{3})\zeta_{12}$, $(1+\sqrt{3})(-1+\zeta_{12})$, $-(1+\sqrt{3})$, $-(1+\sqrt{3})\zeta_{12}$,

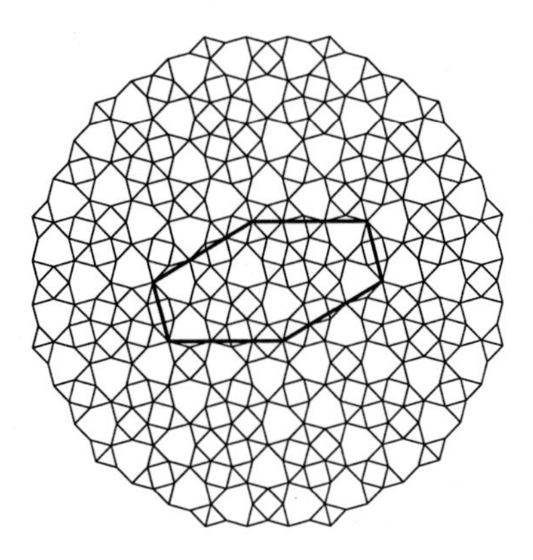

FIGURE 2.5. A central patch of the twelvefold symmetric shield tiling. Therein, one of the affinely regular 6-gons as described in example (S) is shown, the other solutions being rather obvious.

$(1+\sqrt{3})(1-\zeta_{12})$. Alternatively, simply consider the affinely regular polygon R_6 in $\varLambda_{\mathrm{S}}$.

2.3.4.2. *U-polygons in cyclotomic model sets.*

DEFINITION 2.39. For $n \in \mathbb{N}\setminus\{1,2\}$, we denote by $\mathcal{U}^n_{4,\mathbb{Q}}$ the collection of all sets $U \subset \mathbb{S}^1$ of four or more pairwise non-parallel $\mathcal{O}_n$-directions having the property that the cross ratio of slopes of any four directions of U (arranged in arbitrary order) is an element of $\mathbb{Q}$.

LEMMA 2.40. *Let $u \in \mathbb{S}^1$ be an $\mathcal{O}_n$-direction. Then, one has $s_u \in \mathbb{k}_{[n,4]} \cup \{\infty\}$.*

PROOF. Let $u \in \mathbb{S}^1$ be an $\mathcal{O}_n$-direction. By Lemma 2.19 and, since one has $\mathcal{O}_n - \mathcal{O}_n = \mathcal{O}_n$ and $\overline{\mathcal{O}_n} = \mathcal{O}_n$, one obtains

$$s_u \;\in\; \Big(\mathbb{Q}\big((\mathcal{O}_n-\mathcal{O}_n)\cup(\overline{\mathcal{O}_n-\mathcal{O}_n})\cup\{i\}\big)\cap\mathbb{R}\Big)\cup\Big\{\infty\Big\} \;=\; (\mathbb{Q}(\zeta_n,i)\cap\mathbb{R})\cup\{\infty\}\,.$$

The assertion follows from the equality $\mathbb{Q}(\zeta_n,i)=\mathbb{Q}(\zeta_n,\zeta_4)=\mathbb{K}_n\mathbb{K}_4=\mathbb{K}_{[n,4]}$, which implies that $\mathbb{Q}(\zeta_n,i)\cap\mathbb{R}=\mathbb{k}_{[n,4]}$, since $\mathbb{k}_{[n,4]}$ is the maximal real subfield of $\mathbb{K}_{[n,4]}$. □

LEMMA 2.41. *One has the equality $\mathbb{k}_{\mathcal{O}_n}=\mathbb{k}_n$.*

PROOF. Since $\mathcal{O}_n-\mathcal{O}_n=\mathcal{O}_n$ and $\overline{\mathcal{O}_n}=\mathcal{O}_n$, one has

$$\mathbb{k}_{\mathcal{O}_n}=\mathbb{Q}\big((\mathcal{O}_n-\mathcal{O}_n)\cup(\overline{\mathcal{O}_n-\mathcal{O}_n})\big)\cap\mathbb{R} \;=\; \mathbb{Q}(\zeta_n)\cap\mathbb{R}\,.$$

Since $\mathbb{k}_n$ is the maximal real subfield of $\mathbb{K}_n=\mathbb{Q}(\zeta_n)$, the assertion follows. □

In view of Lemma 2.40, it is clear that the cross ratio of slopes of four pairwise non-parallel $\mathcal{O}_n$-directions is an element of the algebraic number field $\mathbb{k}_{[n,4]}$. One even has the following result.

LEMMA 2.42. *The cross ratio of slopes of four pairwise non-parallel $\mathcal{O}_n$-directions is an element of the real algebraic number field $\mathbb{k}_n$.*

PROOF. This follows from Lemma 2.20 in conjunction with Lemma 2.41. □

For $n \in \{3,4,6\}$, the restriction imposed by the definition of the set $\mathcal{U}^n_{4,\mathbb{Q}}$ is always fulfilled.

LEMMA 2.43. *If $n \in \{3,4,6\}$, then $\mathcal{U}^n_{4,\mathbb{Q}}$ is precisely the set of all sets $U \subset \mathbb{S}^1$ of four or more pairwise non-parallel $\mathcal{O}_n$-directions.*

PROOF. Observing that $\mathbb{k}_n = \mathbb{Q}$ for $n \in \{3,4,6\}$, the assertion follows immediately from Lemma 2.42. □

REMARK 2.44. Let $z_j \in \mathcal{O}_4 = \mathbb{Z}^2$, $j \in \{1,\dots,4\}$, be four pairwise non-parallel elements of the square lattice. Clearly, the cross ratio $(s_{z_1}, s_{z_2}; s_{z_3}, s_{z_4})$ is a rational number, say q. Let $n \in \mathbb{N}\setminus\{1,2\}$ and consider the non-singular linear transformation L^n_4 of the Euclidean plane as defined in the proof of Theorem 2.33 (proof of direction (vi) $\Rightarrow$ (i)). By Lemma 2.17, one has

$$q = \left(s_{L^n_4(z_1)}, s_{L^n_4(z_2)}; s_{L^n_4(z_3)}, s_{L^n_4(z_4)}\right).$$

Hence, since L^n_4 maps $\mathbb{Z}^2$ into $\mathcal{O}_n$ (see the proof of Theorem 2.33 (direction (vi) $\Rightarrow$ (i)) or Lemma 1.9(a)), we see that $\mathcal{U}^n_{4,\mathbb{Q}}$ from Definition 2.39 is not empty. In fact, one has

$$V := \left\{u_{L^n_4(z_j)} \,\middle|\, j \in \{1,\dots,4\}\right\} \in \mathcal{U}^n_{4,\mathbb{Q}}.$$

We shall make use of L^n_4 in conjunction with Lemma 1.81 in order to transfer results obtained by Gardner and Gritzmann [**33**] for the square lattice to the class of cyclotomic model sets.

REMARK 2.45. In Remark 2.22, we stated that a U-polygon need not be affinely regular, even if it is a U-polygon in an algebraic Delone set. Now, we are able to give an example in the case of certain cyclotomic model sets.

EXAMPLE 2.46. Let $\varLambda_n(t,W) \in \mathcal{M}(\mathcal{O}_n)$ be cyclotomic model sets with $n \in \{3,4,5,8,12\}$. Thereby, our standard examples $\varLambda_{\mathrm{SQ}}$, $\varLambda_{\mathrm{TRI}}$, $\varLambda_{\mathrm{TTT}}$, $\varLambda_{\mathrm{AB}}$ and $\varLambda_{\mathrm{S}}$ as introduced in Example 1.76 are included. If $n \in \{3,4,5,8\}$, let U consist of the six pairwise non-parallel $\mathcal{O}_n$-directions u_1, $u_{2+\zeta_n}$, $u_{1+\zeta_n}$, $u_{1+2\zeta_n}$, u_{ζ_n} and $u_{-1+\zeta_n}$. Let P be the non-degenerate convex dodecagon with vertices at $3+\zeta_n$, $3+2\zeta_n$, $2+3\zeta_n$, $1+3\zeta_n$, $-1+2\zeta_n$, $-2+\zeta_n$, and the reflections of these points in the origin 0; compare [**33**, Example 4.3 and Figure 1] and see Remark 2.44. Then, one easily verifies that P is a U-polygon in $\mathcal{O}_n$. By Lemma 1.81, there is a homothety $h\colon \mathbb{R}^2 \longrightarrow \mathbb{R}^2$ such that $P' := h[P]$ is a polygon in $\varLambda_n(t,W)$. Since P' is a U-polygon (see Lemma 1.113(a)), P' is a U-polygon in $\varLambda_n(t,W)$. However, by Corollary 2.35(a) and Corollary 2.37, P' is not affinely regular. If $n = 12$, let U consist of the four pairwise non-parallel $\mathcal{O}_{12}$-directions $u_{1-\zeta_{12}}$, u_1, $u_{1+\zeta_{12}}$ and $u_{\zeta_{12}}$. Let P be the non-degenerate convex octagon with vertices at 1, ζ_{12}, $-1+\zeta_{12}$, -2, $-2-\zeta_{12}$, $-1-2\zeta_{12}$, $-2\zeta_{12}$ and $1-\zeta_{12}$; compare [**41**, Ch. 4, Figure 4.3] and see Remark 2.44. Then, one easily verifies that P is a U-polygon in $\mathcal{O}_{12}$. By Lemma 1.81, there is a homothety $h\colon \mathbb{R}^2 \longrightarrow \mathbb{R}^2$ such that $P' := h[P]$ is a polygon in $\varLambda_{12}(t,W)$. Since P' is a U-polygon (see Lemma 1.113(a)), P' is a U-polygon in $\varLambda_{12}(t,W)$. However, by Corollary 2.37, P' is not affinely regular.

The proof of the following result is step by step analogous to that of [**33**, Theorem 4.5], wherefore we need not repeat it here. The most important tools for this proof are Proposition 2.21 and p-adic valuations.

THEOREM 2.47. *Let $n \in \mathbb{N} \setminus \{1,2\}$, let $U \in \mathcal{U}^n_{4,\mathbb{Q}}$, and suppose the existence of a U-polygon. Then, one has* $\mathrm{card}(U) \leq 6$, *and the cross ratio of slopes of any four directions of U, arranged in order of increasing angle with the positive real axis, is an element of the set N_1 (as defined in Theorem 1.40).* □

REMARK 2.48. Due to the present state of our analysis in Section 1.1.5, we are presently unable to prove a full analogue of Theorem 2.47 in this section in the case of arbitrary sets of four or more pairwise non-parallel $\mathcal{O}_n$-directions. In fact, it is an open problem whether there really always is a cardinality statement as above.

THEOREM 2.49. *Let $n \in \mathbb{N} \setminus \{1,2\}$, let $U \subset \mathbb{S}^1$ be a set of four or more pairwise non-parallel $\mathcal{O}_n$-directions, and suppose the existence of a U-polygon. Then, the cross ratio of slopes of any four directions of U, arranged in order of increasing angle with the positive real axis, is an element of the set*

$$\Big(\bigcup_{m \geq 4} f_m[D_m] \Big) \cap \mathbb{k}_n \,,$$

with D_m and f_m as in Definition 1.37.

PROOF. This follows from Theorem 2.24 in conjunction with Lemma 2.41. □

Suppose that there is a U-polygon in an (aperiodic) cyclotomic model set with codimension two. Then, the set U consists of $\mathcal{O}_n$-directions, where $n \in \{5, 8, 10, 12\}$.

THEOREM 2.50. *Let $n \in \{5, 8, 10, 12\}$, let $U \subset \mathbb{S}^1$ be a set of four or more pairwise non-parallel $\mathcal{O}_n$-directions, and suppose the existence of a U-polygon. Then, the cross ratio of slopes of any four directions of U, arranged in order of increasing angle with the positive real axis, maps under the norm $N_{\mathbb{k}_n/\mathbb{Q}}$ to the set N_2 as defined in Theorem 1.48.*

PROOF. This follows from Theorem 2.49 in conjunction with Corollary 1.53. □

For the general case $n \in \mathbb{N} \setminus \{1,2\}$, one obtains the following results.

THEOREM 2.51. *For all $e \in \mathrm{Im}(\frac{\phi}{2})$, there is a finite set $N_e \subset \mathbb{Q}$ such that, for all $n \in (\frac{\phi}{2})^{-1}[\{e\}]$, and all sets $U \subset \mathbb{S}^1$ of four or more pairwise non-parallel $\mathcal{O}_n$-directions, one has the following:*

If there exists a U-polygon, then the cross ratio of slopes of any four directions of U, arranged in order of increasing angle with the positive real axis, maps under the norm $N_{\mathbb{k}_n/\mathbb{Q}}$ to N_e.

PROOF. This follows from Theorem 2.49 in conjunction with Corollary 1.56. □

THEOREM 2.52. *For all $n \in \mathbb{N} \setminus \{1,2\}$ and all sets $U \subset \mathbb{S}^1$ of four or more pairwise non-parallel $\mathcal{O}_n$-directions, one has the following:*

If there exists a U-polygon, then the cross ratio of slopes of any four directions of U, arranged in order of increasing angle with the positive real axis, maps under the norm $N_{\mathbb{k}_n/\mathbb{Q}}$ to the set

$$\pm \left(\{1\} \cup \left[\mathbb{Q}_{>1}^{\mathbb{P}_{\leq 2}}\right]^{\pm 1} \right).$$

PROOF. This follows from Theorem 2.49 in conjunction with Theorem 1.51. □

2.3.4.3. *Determination of convex subsets of cyclotomic model sets by X-rays.*

THEOREM 2.53. *Let* $n \in \mathbb{N} \setminus \{1, 2\}$ *and let* $\Lambda_n(t, W) \in \mathcal{M}(\mathcal{O}_n)$ *be a cyclotomic model set. Further, let* $U \subset \mathbb{S}^1$ *be a set of two or more pairwise non-parallel* $\mathcal{O}_n$*-directions. The following statements are equivalent:*

(i) $\mathcal{C}(\Lambda_n(t, W))$ *is determined by the X-rays in the directions of U.*
(ii) *There is no U-polygon in* $\Lambda_n(t, W)$.

PROOF. This follows immediately from Theorem 2.29 in conjunction with Lemma 1.83 and Lemma 1.84(a). □

We are now able to prove the following result.

THEOREM 2.54. *Let* $n \in \mathbb{N} \setminus \{1, 2\}$. *Then, one has:*

(a) *There is a set* $U \in \mathcal{U}^n_{4,\mathbb{Q}}$ *with* $\operatorname{card}(U) = 4$ *such that, for all cyclotomic model sets* $\Lambda_n(t, W) \in \mathcal{M}(\mathcal{O}_n)$, *the set* $\mathcal{C}(\Lambda_n(t, W))$ *is determined by the X-rays in the directions of U. Furthermore, any set* $U \in \mathcal{U}^n_{4,\mathbb{Q}}$ *with* $\operatorname{card}(U) = 4$ *having the property that the cross ratio of slopes of the directions of U, arranged in order of increasing angle with the positive real axis, is not an element of the set* N_1 *(as defined in Theorem 1.40) is suitable for this purpose.*
(b) *Let* $U \subset \mathbb{S}^1$ *be any set of three or less pairwise non-parallel* $\mathcal{O}_n$*-directions. Then, for all cyclotomic model sets* $\Lambda_n(t, W) \in \mathcal{M}(\mathcal{O}_n)$, $\mathcal{C}(\Lambda_n(t, W))$ *is not determined by the X-rays in the directions of U.*
(c) *Let* $U \in \mathcal{U}^n_{4,\mathbb{Q}}$ *with* $\operatorname{card}(U) = 7$. *Then, for all cyclotomic model sets* $\Lambda_n(t, W) \in \mathcal{M}(\mathcal{O}_n)$, *the set* $\mathcal{C}(\Lambda_n(t, W))$ *is determined by the X-rays in the directions of U.*
(d) *There is a set* $U \in \mathcal{U}^n_{4,\mathbb{Q}}$ *with* $\operatorname{card}(U) = 6$ *such that, for all cyclotomic model sets* $\Lambda_n(t, W) \in \mathcal{M}(\mathcal{O}_n)$, *the set* $\mathcal{C}(\Lambda_n(t, W))$ *is not determined by the X-rays in the directions of U.*

PROOF. Let us start with (a). In view of Theorem 2.47 and Theorem 2.53, the additional statement is immediate. So, it remains to show existence. For example, the following subsets of $\mathcal{U}^n_{4,\mathbb{Q}}$ have the property that the cross ratio of slopes of their directions, arranged in order of increasing angle with the positive real axis, is not an element of the set N_1:

$$U'_n := \{u_1, u_{1+\zeta_n}, u_{1+2\zeta_n}, u_{1+5\zeta_n}\}, \quad U''_n := \{u_1, u_{2+\zeta_n}, u_{\zeta_n}, u_{-1+2\zeta_n}\}$$

and $U'''_n := \{u_{2+\zeta_n}, u_{3+2\zeta_n}, u_{1+\zeta_n}, u_{2+3\zeta_n}\}$; compare [**33**, Remark 5.8] and Remark 2.44. For these sets, the cross ratio of slopes of the directions, arranged in order of increasing angle with the positive real axis, is, in order of their appearance above, 8/5, 5/4 and 5/4 again.

Assertions (b) and (c) are immediate consequences of Theorem 2.53 in conjunction with Lemma 2.23, Lemma 1.83 and Theorem 2.47, respectively.

For (d), note that there is a U-polygon P' in $\Lambda_n(t, W)$, where $U \in \mathcal{U}^n_{4,\mathbb{Q}}$ consists of the six pairwise non-parallel $\mathcal{O}_n$-directions u_1, $u_{2+\zeta_n}$, $u_{1+\zeta_n}$, $u_{1+2\zeta_n}$, u_{ζ_n} and $u_{-1+\zeta_n}$, respectively. To see this, let P be the non-degenerate convex dodecagon with vertices at $3 + \zeta_n$, $3 + 2\zeta_n$, $2 + 3\zeta_n$, $1 + 3\zeta_n$, $-1 + 2\zeta_n$, $-2 + \zeta_n$, and the reflections of these points in the origin 0; compare [**33**, Example 4.3 and Figure 1] and see Remark 2.44. Then, one easily verifies that P is a U-polygon in $\mathcal{O}_n$. By Lemma 1.81, there is a homothety $h\colon \mathbb{R}^2 \longrightarrow \mathbb{R}^2$ such that

$P' := h[P]$ is a polygon in $\varLambda_n(t,W)$. Since P' is a U-polygon (see Lemma 1.113(a)), P' is a U-polygon in $\varLambda_n(t,W)$. The assertion now follows immediately from Theorem 2.53. □

REMARK 2.55. By assertions (b) and (d) of Theorem 2.54, assertions (a) and (c) of Theorem 2.54 are optimal with respect to the number of directions used.

As shown in Lemma 2.43 for $n \in \{3,4,6\}$ (corresponding to periodic cyclotomic model sets), the set $\mathcal{U}^n_{4,\mathbb{Q}}$ coincides with the natural set of all sets $U \subset \mathbb{S}^1$ of four or more pairwise non-parallel $\mathcal{O}_n$-directions. Though, for $n \in \mathbb{N} \setminus \{3,4,6\}$, the restriction of the set of all sets of four or more pairwise non-parallel $\mathcal{O}_n$-directions to the set $\mathcal{U}^n_{4,\mathbb{Q}}$ in Theorem 2.54 is rather artificial. Now, we use our results from Section 2.3.4.2 in order to remove this restriction and prove corresponding more results in larger generality. For the case $n \in \{5,8,10,12\}$ (corresponding to (aperiodic) cyclotomic model sets $\varLambda_n(t,W) \in \mathcal{M}(\mathcal{O}_n)$ with co-dimension two), one has the following generalization of Theorem 2.54(a).

THEOREM 2.56. *Let $n \in \{5,8,10,12\}$ and let $U \subset \mathbb{S}^1$ be any set of four pairwise non-parallel $\mathcal{O}_n$-directions having the property*

(C2) *The cross ratio of slopes of the directions of U, arranged in order of increasing angle with the positive real axis, does not map under the norm $N_{\mathbb{k}_n/\mathbb{Q}}$ to the set N_2 as defined in Theorem 1.48.*

Then, for all (aperiodic) cyclotomic model sets $\varLambda_n(t,W) \in \mathcal{M}(\mathcal{O}_n)$, $\mathcal{C}(\varLambda_n(t,W))$ is determined by the X-rays in the directions of U.

PROOF. This follows immediately from Theorem 2.53 and Theorem 2.50. □

EXAMPLE 2.57. Let us present suitable examples for the practically relevant cases. As an example with $n = 8$, the following set of $\mathcal{O}_8$-directions has property (C2):

$$U_8 := \left\{u_{1+\zeta_8}, u_{(-1+\sqrt{2})+\sqrt{2}\zeta_8}, u_{(-1-\sqrt{2})+\zeta_8}, u_{-2+(-1+\sqrt{2})\zeta_8}\right\}.$$

Here, the cross ratio of slopes of the elements of U_8, arranged in order of increasing angle with the positive real axis, equals $\frac{12}{7} - \frac{3}{7}\sqrt{2}$, hence $N_{\mathbb{k}_8/\mathbb{Q}}(\frac{12}{7} - \frac{3}{7}\sqrt{2}) = \frac{18}{7} \notin N_2$. Further, for $n \in \{5,10\}$, the following set of $\mathcal{O}_5$-directions has property (C2):

$$U_{10} := U_5 := \left\{u_{(1+\tau)+\zeta_5}, u_{(\tau-1)+\zeta_5}, u_{-\tau+\zeta_5}, u_{2\tau-\zeta_5}\right\}.$$

Here, the cross ratio of slopes of the elements of U_5, arranged in order of increasing angle with the positive real axis, equals $\frac{4}{5} + \frac{1}{5}\sqrt{5}$, hence $N_{\mathbb{k}_5/\mathbb{Q}}(\frac{4}{5} + \frac{1}{5}\sqrt{5}) = \frac{11}{25} \notin N_2$. Finally, for $n = 12$, the following set of $\mathcal{O}_{12}$-directions has property (C2):

$$U_{12} := \left\{u_1, u_{2+\zeta_{12}}, u_{\zeta_{12}}, u_{\sqrt{3}-\zeta_{12}}\right\}.$$

Here, the cross ratio of slopes of the elements of U_{12}, arranged in order of increasing angle with the positive real axis, equals $2 + \frac{1}{2}\sqrt{3}$, hence $N_{\mathbb{k}_{12}/\mathbb{Q}}(2 + \frac{1}{2}\sqrt{3}) = \frac{13}{4} \notin N_2$.

REMARK 2.58. Using the fact that the quadratic fields $\mathbb{k}_8, \mathbb{k}_5$ and $\mathbb{k}_{12}$ have class number one, Pleasants [**52**] showed that the sets of $\mathcal{O}_n$-directions U_8, U_5 and U_{12} in Example 2.57 are well suited in order to yield dense lines in the corresponding (aperiodic) cyclotomic model sets. Note that precise statements on densities of lines in aperiodic cyclotomic model sets depend on the shape of the window. It follows that, in the practice of quantitative HRTEM, the resolution coming from the above directions is likely to be rather high, which makes Theorem 2.56 look promising.

For the general case $n \in \mathbb{N}\setminus\{1,2\}$, one has the following generalizations of Theorem 2.54(a) and Theorem 2.56, dealing with arbitrary (but fixed) cyclotomic model sets and arbitrary sets U of four pairwise non-parallel $\mathcal{O}_n$-directions.

THEOREM 2.59. *For all $e \in \mathrm{Im}(\frac{\phi}{2})$, there is a finite set $N_e \subset \mathbb{Q}$ such that, for all $n \in (\frac{\phi}{2})^{-1}[\{e\}]$, and all sets $U \subset \mathbb{S}^1$ of four pairwise non-parallel $\mathcal{O}_n$-directions, one has the following:*

If U has the property that the cross ratio of slopes of the directions of U, arranged in order of increasing angle with the positive real axis, does not map under the norm $N_{\mathbb{k}_n/\mathbb{Q}}$ to N_e, then, for all cyclotomic model sets $\varLambda_n(t,W) \in \mathcal{M}(\mathcal{O}_n)$, the set $\mathcal{C}(\varLambda_n(t,W))$ is determined by the X-rays in the directions of U.

PROOF. This follows immediately from Theorem 2.53 and Theorem 2.51. □

THEOREM 2.60. *For all $n \in \mathbb{N}\setminus\{1,2\}$ and all sets $U \subset \mathbb{S}^1$ of four pairwise non-parallel $\mathcal{O}_n$-directions, one has the following:*

If U has the property that the cross ratio of slopes of the directions of U, arranged in order of increasing angle with the positive real axis, does not map under the norm $N_{\mathbb{k}_n/\mathbb{Q}}$ to the set

$$\pm\left(\{1\} \cup \left[\mathbb{Q}_{>1}^{\mathbb{P}_{\leq 2}}\right]^{\pm 1}\right),$$

then, for all cyclotomic model sets $\varLambda_n(t,W) \in \mathcal{M}(\mathcal{O}_n)$, the set $\mathcal{C}(\varLambda_n(t,W))$ is determined by the X-rays in the directions of U.

PROOF. This follows immediately from Theorem 2.53 and Theorem 2.52. □

For the case of periodic cyclotomic model sets, we are able to prove the first uniqueness result, which is in full accordance with the setting described in Section 1.3.

THEOREM 2.61. *Let $n \in \{3,4,6\}$ and let $U \subset \mathbb{S}^1$ be any set of four pairwise non-parallel $\mathcal{O}_n$-directions having the following properties.*

(C1) *The cross ratio of slopes of the directions of U, arranged in order of increasing angle with the positive real axis, is not an element of the set N_1 as defined in Theorem 1.40.*

(E) *U contains two $\mathcal{O}_n$-directions of the form $u_o, u_{o'}$, where $o, o' \in \mathcal{O}_n \setminus \{0\}$ satisfy one of the equivalent conditions (i)-(iii) of Proposition 1.129.*

Then, the set $\bigcup_{t\in\mathbb{R}^2} \mathcal{C}(t+\mathcal{O}_n)$ is determined by the X-rays in the directions of U.

PROOF. Let $n \in \{3,4,6\}$ and let $U \subset \mathbb{S}^1$ be a set of four pairwise non-parallel $\mathcal{O}_n$-directions having the Properties (C1) and (E). Let $F, F' \in \bigcup_{t\in\mathbb{R}^2} \mathcal{C}(t+\mathcal{O}_n)$, say $F \in \mathcal{C}(t+\mathcal{O}_n)$ and $F' \in \mathcal{C}(t'+\mathcal{O}_n)$, where $t, t' \in \mathbb{R}^2$, and suppose that F and F' have the same X-rays in the directions of U. Then, by Lemma 1.120 and Theorem 1.130 in conjunction with property (E), one obtains

$$F, F' \subset G_U^F \subset t + \mathcal{O}_n\,. \tag{2.3}$$

If $F = \varnothing$, then, by Lemma 1.112(a), one also gets $F' = \varnothing$. It follows that one may assume, without loss of generality, that F and F' are non-empty. Then, since $F' \subset t' + \mathcal{O}_n$, it follows from Relation (2.3) that $t + \mathcal{O}_n$ meets $t' + \mathcal{O}_n$, the latter being equivalent to the identity $t + \mathcal{O}_n = t' + \mathcal{O}_n$. Now, the assertion follows immediately from property (C1) and Theorem 2.54(a) in conjunction with Lemma 2.43. □

EXAMPLE 2.62. It was shown in the proof of Theorem 2.54(a) that, for $n \in \{3,4\}$, the sets of $\mathcal{O}_n$-directions U'_n, U''_n, U'''_n (as defined in the proof of Theorem 2.54(a)) have property (C1). One can easily see that these sets of $\mathcal{O}_n$-directions additionally have property (E).

REMARK 2.63. For $n \in \{3,4\}$, the sets of $\mathcal{O}_n$-directions parallel to the elements of the sets U'_n, U''_n, U'''_n (as defined in the proof of Theorem 2.54(a)) obviously yield dense lines in the corresponding (periodic) cyclotomic model sets. It follows that, in the practice of quantitative HRTEM, the resolution coming from these directions is rather high, so there might be a real application of Theorem 2.61.

REMARK 2.64. In an approximative sense, which will be made precise below, one is also able to deal with the 'non-anchored' case for regular aperiodic cyclotomic model sets, i.e., the determination of sets of the form $\bigcup_{\Lambda \in \mathcal{M}_g^{\star_n}(W)} \mathcal{C}(\Lambda)$, where $n \in \mathbb{N} \setminus \{1,2,3,4,6\}$, by the X-rays in four prescribed $\mathcal{O}_n$-directions. Note first that, for $n \in \mathbb{N} \setminus \{1,2,3,4,6\}$, Theorem 2.59 shows that there is a finite set $N_{\phi(n)/2} \subset \mathbb{Q}$ such that, for all sets $U \subset \mathbb{S}^1$ of four pairwise non-parallel $\mathcal{O}_n$-directions, one has:

If U has the property that the cross ratio of slopes of the directions of U, arranged in order of increasing angle with the positive real axis, does not map under the norm $N_{\mathbb{k}_n/\mathbb{Q}}$ to $N_{\phi(n)/2}$, then, for all cyclotomic model sets $\Lambda_n(t,W) \in \mathcal{M}(\mathcal{O}_n)$, the set $\mathcal{C}(\Lambda_n(t,W))$ is determined by the X-rays in the directions of U.

Now let $U \subset \mathbb{S}^1$ be any set of four pairwise non-parallel $\mathcal{O}_n$-directions having the following properties.

- (C) The cross ratio of slopes of the directions of U, arranged in order of increasing angle with the positive real axis, does not map under the norm $N_{\mathbb{k}_n/\mathbb{Q}}$ to the set $N_{\phi(n)/2}$ from above.
- (E) U contains two $\mathcal{O}_n$-directions of the form $u_o, u_{o'}$, where $o, o' \in \mathcal{O}_n \setminus \{0\}$ satisfy one of the equivalent conditions (i)-(iii) of Proposition 1.129.

We claim that, in a sense, for all fixed windows $W \subset (\mathbb{R}^2)^{\phi(n)/2-1}$ with boundary $\mathrm{bd}(W)$ having Lebesgue measure 0 in $(\mathbb{R}^2)^{\phi(n)/2-1}$ and all fixed star maps

$$.^{\star_n} : \mathcal{O}_n \longrightarrow (\mathbb{R}^2)^{\frac{\phi(n)}{2}-1}$$

(as described in Definition 1.73), the set $\bigcup_{\Lambda \in \mathcal{M}_g^{\star_n}(W)} \mathcal{C}(\Lambda)$ is determined by the X-rays in the directions of U. To see this, let $F, F' \in \bigcup_{\Lambda \in \mathcal{M}_g^{\star_n}(W)} \mathcal{C}(\Lambda)$, say $F \in \mathcal{C}(\Lambda_n^{\star_n}(t, \tau + W))$ and $F' \in \mathcal{C}(\Lambda_n^{\star_n}(t', \tau' + W))$, where $t, t' \in \mathbb{R}^2$ and $\tau, \tau' \in (\mathbb{R}^2)^{\frac{\phi(n)}{2}-1}$, and suppose that F and F' have the same X-rays in the directions of U. Then, by Lemma 1.120 and Theorem 1.130 in conjunction with property (E), one obtains

$$F, F' \subset G_U^F \subset t + \mathcal{O}_n\,. \tag{2.4}$$

If $F = \varnothing$, then, by Lemma 1.112(a), one also gets $F' = \varnothing$. One may thus assume, without loss of generality, that F and F' are non-empty. Then, since $F' \subset t' + \mathcal{O}_n$, Relation (2.4) implies that $t + \mathcal{O}_n$ meets $t' + \mathcal{O}_n$, the latter being equivalent to the identity $t + \mathcal{O}_n = t' + \mathcal{O}_n$.

Moreover, the identity $t + \mathcal{O}_n = t' + \mathcal{O}_n$ is equivalent to the relation $t' - t \in \mathcal{O}_n$. Hence, one has

$$F - t \in \mathcal{C}\Big(\Lambda_n^{\star n}(0, \tau + W)\Big)$$

and, since the equality

$$\Lambda_n^{\star n}\Big(t' - t, \tau' + W\Big) = \Lambda_n^{\star n}\Big(0, (\tau' + (t' - t)^{\star n}) + W\Big)$$

holds,

$$F' - t \in \mathcal{C}\Big(\Lambda_n^{\star n}(t' - t, \tau' + W)\Big) = \mathcal{C}\Big(\Lambda_n^{\star n}(0, (\tau' + (t' - t)^{\star n}) + W)\Big).$$

Clearly, $F - t$ and $F' - t$ again have the same X-rays in the directions of U. Hence, by Lemma 1.112(b), $F - t$ and $F' - t$ have the same centroid. Since the star map $.^{\star n}$ is $\mathbb{Q}$-linear, it follows that the finite subsets $[F - t]^{\star n}$ and $[F' - t]^{\star n}$ of $(\mathbb{R}^2)^{\phi(n)/2-1}$ also have the same centroid. Now, if one has

$$F - t = B_R(a) \,\cap\, \Lambda_n^{\star n}(t, \tau + W)$$

and

$$F' - t = B_{R'}(a') \,\cap\, \Lambda_n^{\star n}(0, (\tau' + (t' - t)^{\star n}) + W)$$

for suitable $a, a' \in \mathbb{R}^2$ and large $R, R' > 0$ (which is rather natural in practice), then Theorem 1.90 allows us to write

$$\begin{aligned}
\frac{1}{\mathrm{vol}(W)} \int_{\tau + W} y \, \mathrm{d}\lambda(y) &\approx \frac{1}{\mathrm{card}\,(F - t)} \sum_{x \in F - t} x^{\star n} \\
&= \frac{1}{\mathrm{card}\,(F' - t)} \sum_{x \in F' - t} x^{\star n} \\
&\approx \frac{1}{\mathrm{vol}(W)} \int_{(\tau' + (t' - t)^{\star n}) + W} y \, \mathrm{d}\lambda(y)\,.
\end{aligned}$$

Consequently,

$$\tau + \int_W y \, \mathrm{d}\lambda(y) \approx (\tau' + (t' - t)^{\star n}) + \int_W y \, \mathrm{d}\lambda(y)\,,$$

and hence $\tau \approx \tau' + (t' - t)^{\star n}$. The latter means that, approximately, both $F - t$ and $F' - t$ are elements of the set $\mathcal{C}(\Lambda_n^{\star n}(0, \tau + W))$. Now, it follows in this approximative sense from property (C) and Theorem 2.59 that $F - t \approx F' - t$, and hence $F \approx F'$.

EXAMPLE 2.65. For $n \in \{5, 8, 10, 12\}$, the set of $\mathcal{O}_n$-direction U_n (as defined in Example 2.57) has property (C) with N_2 as defined in Theorem 1.48. One can easily see that these $\mathcal{O}_n$-directions additionally have property (E).

2.3.4.4. *Determination of convex subsets of icosahedral model sets by X-rays.* Although an example in 3-space, the sliced structure of icosahedral model sets permits a lift of our previous results to this situation.

REMARK 2.66. Note that, for a convex subset C of an icosahedral model set $\Lambda_{\mathrm{ico}}(t, W)$, the intersections $C \cap (\lambda + H^{(\tau,0,1)})$, $\lambda \in \Lambda_{\mathrm{ico}}(t, W)$, are convex subsets of $\Lambda_{\mathrm{ico}}(t, W) \cap (\lambda + H^{(\tau,0,1)})$, i.e., they satisfy

$$\mathrm{conv}\Big(C \cap \Big(\lambda + H^{(\tau,0,1)}\Big)\Big) \cap \Big(\Lambda_{\mathrm{ico}}(t, W) \cap \Big(\lambda + H^{(\tau,0,1)}\Big)\Big) = C \cap \Big(\lambda + H^{(\tau,0,1)}\Big)\,.$$

THEOREM 2.67. *There is a set $U \subset \mathbb{S}^2$ of four $\mathrm{Im}[\mathbb{I}]^{(\tau,0,1)}$-directions such that, for all icosahedral model sets $\Lambda_{\mathrm{ico}}(t,W)$, the set $\mathcal{C}(\Lambda_{\mathrm{ico}}(t,W))$ is determined by the X-rays in the directions of U.*

PROOF. By Remark 2.66, Remark 1.98 in conjunction with Corollary 1.102, this follows immediately by applying Theorem 2.56 to each 'slice'

$$\Phi\left[\left(\Lambda_{\mathrm{ico}}(t,W)\cap\left(\lambda+H^{(\tau,0,1)}\right)\right)-\lambda\right],$$

where $\lambda \in \Lambda_{\mathrm{ico}}(t,W)$. For an example, it was shown in Example 2.57 that the convex subsets of cyclotomic model sets with underlying $\mathbb{Z}$-module $\mathcal{O}_5$ are determined by the X-rays in the $\mathcal{O}_5$-directions given by $U_5 := \{u_{(1+\tau)+\zeta_5}, u_{(\tau-1)+\zeta_5}, u_{-\tau+\zeta_5}, u_{2\tau-\zeta_5}\}$. Consequently, the convex subsets of icosahedral model sets $\Lambda_{\mathrm{ico}}(t,W)$ are determined by the X-rays in the $\mathrm{Im}[\mathbb{I}]^{(\tau,0,1)}$-directions given by

$$U_{\mathrm{ico}} := \Phi^{-1}[U_5] = \left\{u_{(-\frac{1}{2},\frac{1}{2}+\frac{3}{2}\tau,\frac{1}{2}\tau)^t}, u_{(-\frac{1}{2},-\frac{3}{2}+\frac{3}{2}\tau,\frac{1}{2}\tau)^t}, u_{(-\frac{1}{2},-\frac{1}{2}-\frac{1}{2}\tau,\frac{1}{2}\tau)^t}, u_{(\frac{1}{2},\frac{1}{2}+\frac{3}{2}\tau,-\frac{1}{2}\tau)^t}\right\}.$$

This completes the proof. □

REMARK 2.68. The proof of Theorem 2.67 shows that the result even extends to the set of subsets C of icosahedral model sets $\Lambda_{\mathrm{ico}}(t,W)$ that are only $H^{(\tau,0,1)}$-*convex*, the latter meaning that, for all $\lambda \in \Lambda_{\mathrm{ico}}(t,W)$, the sets $C\cap(\lambda+H^{(\tau,0,1)})$ are convex subsets of $\Lambda_{\mathrm{ico}}(t,W)\cap(\lambda+H^{(\tau,0,1)})$. Since, by Remark 2.58, the set of $\mathcal{O}_5$-directions U_5 (as defined in the proof of Theorem 2.67) is well suited in order to yield dense lines in cyclotomic model sets with underlying $\mathbb{Z}$-module $\mathcal{O}_5$, it follows from the results of Section 1.2.3.3 that the set of $\mathrm{Im}[\mathbb{I}]^{(\tau,0,1)}$-directions U_{ico} (as defined in the proof of Theorem 2.67) is well suited in order to yield dense lines in the corresponding slices of icosahedral model sets $\Lambda_{\mathrm{ico}}(t,W)$, i.e., sets of the form

$$\Lambda_{\mathrm{ico}}(t,W)\cap\left(\lambda+H^{(\tau,0,1)}\right),\quad \lambda\in\Lambda_{\mathrm{ico}}(t,W). \tag{2.5}$$

In fact, these directions even yield dense lines in icosahedral model sets as a whole; see **[52]** again. In particular, neighbouring slices of the form (2.5) are densely occupied and hence well separated. Consequently, neighbouring lines in any of the directions of U_{ico} that meet at least one point of a fixed icosahedral model set $\Lambda_{\mathrm{ico}}(t,W)$ are sufficiently separated. It follows that, in the practice of quantitative HRTEM, the resolution coming from the above directions is likely to be rather high, which makes Theorem 2.67 look promising.

Moreover, in an approximative sense (analogous to the one in Remark 2.64), Theorem 2.67 even holds in the 'non-anchored' case for regular generic icosahedral model sets, i.e., the sets of the form $\bigcup_{\Lambda\in\mathcal{I}_g(W)}\mathcal{C}(\Lambda)$ are approximately determined by the X-rays in four prescribed $\mathrm{Im}[\mathbb{I}]^{(\tau,0,1)}$-directions (e.g., by the X-rays in the directions of U_{ico}); compare Remark 2.64.

Similar to the proof of Theorem 2.67, the following result follows immediately from Theorem 2.54 in conjunction with Remark 2.66, Remark 1.98 and Corollary 1.102.

THEOREM 2.69. *The following statements hold:*

(a) *There is a set $U \in \mathcal{U}^5_{4,\mathbb{Q}}$ with $\mathrm{card}(U) = 4$ such that, for all icosahedral model sets $\Lambda_{\mathrm{ico}}(t,W)$, the set $\mathcal{C}(\Lambda_{\mathrm{ico}}(t,W))$ is determined by the X-rays in the directions of $\Phi^{-1}[U]$. Furthermore, any set $U \in \mathcal{U}^5_{4,\mathbb{Q}}$ with $\mathrm{card}(U) = 4$ having the property* *that the cross ratio of slopes of the directions of U, arranged in order of increasing angle with the positive real axis, is not an element of the set N_1 (as defined in Theorem 1.40) is suitable for this purpose.*

(b) *Let $U \subset \mathbb{S}^1$ be any set of three or less pairwise non-parallel $\mathcal{O}_5$-directions. Then, for all icosahedral model sets $\Lambda_{\text{ico}}(t,W)$, the set $\mathcal{C}(\Lambda_{\text{ico}}(t,W))$ is not determined by the X-rays in the directions of $\Phi^{-1}[U]$.*

(c) *Let $U \in \mathcal{U}^5_{4,\mathbb{Q}}$ with $\text{card}(U) = 7$. Then, for all icosahedral model sets $\Lambda_{\text{ico}}(t,W)$, the set $\mathcal{C}(\Lambda_{\text{ico}}(t,W))$ is determined by the X-rays in the directions of $\Phi^{-1}[U]$.*

(d) *There is a set $U \in \mathcal{U}^5_{4,\mathbb{Q}}$ with $\text{card}(U) = 6$ such that, for all icosahedral model sets $\Lambda_{\text{ico}}(t,W)$, the set $\mathcal{C}(\Lambda_{\text{ico}}(t,W))$ is not determined by the X-rays in the directions of $\Phi^{-1}[U]$.* □

REMARK 2.70. Note that, in the situation of Theorem 2.69, the sets of the form $\Phi^{-1}[U]$ are contained in the set of $\text{Im}[\mathbb{I}]^{(\tau,0,1)}$-directions.

2.4. Successive determination of finite sets

Here, we shall investigate the successive determination of arbitrary finite subsets (not necessarily convex) of arbitrary Delone sets living on $\mathcal{O}_n$, where $n \geq 3$, with applications to cyclotomic and icosahedral model sets. Recall that the interactive technique of successive determination allows us to use the information from previous X-rays in deciding on the direction for the next X-ray.

Although this section will be more technical, the guiding idea of our proof is rather simple and looks as follows. Namely, using Minkowski representations of orders of algebraic number fields, we shall prove the generalization of the following fact to all rings $\mathcal{O}_n$ of cyclotomic integers, where $n \in \mathbb{N} \setminus \{1,2\}$. That is, given $n \in \{3,4,6\}$ (corresponding to the crystallographic cases of the plane), a finite subset F of the (full) lattice $\mathcal{O}_n$ in $\mathbb{R}^2$ and an $\mathcal{O}_n$-direction $u \in \mathbb{S}^1$, there is an $\mathcal{O}_n$-direction $u' \in \mathbb{S}^1$ such that each line in the plane parallel to u' meets at most one point of the set $G^F_{\{u\}} \cap \mathcal{O}_n$.

2.4.1. Lines in rings of cyclotomic integers.

DEFINITION 2.71. For $n \in \mathbb{N} \setminus \{1,2\}$ and $j \in \{1,\dots,\phi(n)/2\}$, set $b^{(n)}_j := (\zeta_n + \bar{\zeta}_n)^{j-1}$ and $b^{(n)}_{\phi(n)/2+j} := (\zeta_n + \bar{\zeta}_n)^{j-1}\zeta_n$. Further, set

$$\begin{aligned} B^{(n)}_1 &:= \{b^{(n)}_1,\dots,b^{(n)}_{\phi(n)/2}\},\\ B^{(n)}_2 &:= \{b^{(n)}_{\phi(n)/2+1},\dots,b^{(n)}_{\phi(n)}\},\end{aligned}$$

and, finally, $B^{(n)} := B^{(n)}_1 \,\dot{\cup}\, B^{(n)}_2$.

LEMMA 2.72. *Let $n \in \mathbb{N}\setminus\{1,2\}$. Then, the set $B^{(n)}$ is both a $\mathbb{Q}$-basis of $\mathbb{K}_n$ and a $\mathbb{Z}$-basis of $\mathcal{O}_n$.*

PROOF. The assertion follows immediately from Lemma 1.9, Corollary 1.14 and the second part of Remark 1.24. □

DEFINITION 2.73. For $n \in \mathbb{N} \setminus \{1,2\}$, let $\,.^{\sim n}$ be an arbitrary, but fixed Minkowski embedding of $\mathbb{K}_n$, i.e., a map $\,.^{\sim n} : \mathbb{K}_n \longrightarrow (\mathbb{R}^2)^{\phi(n)/2}$, given by

$$z \longmapsto \left(\sigma_1(z), \sigma_2(z), \dots, \sigma_{\frac{\phi(n)}{2}}(z)\right),$$

where the set $\{\sigma_1, \dots, \sigma_{\phi(n)/2}\}$ arises from $G(\mathbb{K}_n/\mathbb{Q})$ by choosing exactly one automorphism from each pair of complex conjugate ones.

REMARK 2.74. For the notion of Minkowski embeddings, compare also Remark 1.72 and references given there. Note that the map $\,.^{\sim n}$ is $\mathbb{Q}$-linear and injective. Further, by Lemma 2.72, the set $[\mathcal{O}_n]^{\sim n}$ is a lattice in $(\mathbb{R}^2)^{\phi(n)/2}$ with basis $[B^{(n)}]^{\sim n}$; see [**18**, Ch. 2, Sec. 3]. Further, again by Lemma 2.72 and the $\mathbb{Q}$-linearity of $\,.^{\sim n}$, one has $[\mathbb{K}_n]^{\sim n} = \langle [B^{(n)}]^{\sim n} \rangle_{\mathbb{Q}}$.

DEFINITION 2.75. For $n \in \mathbb{N}\setminus\{1,2\}$, we let $\,.^{\frown n}$ be the co-restriction of $\,.^{\sim n}$ to its image $[\mathbb{K}_n]^{\sim n}$.

LEMMA 2.76. *Let $n \in \mathbb{N} \setminus \{1,2\}$. Then, $\,.^{\frown n}$ is a $\mathbb{Q}$-linear isomorphism.*

PROOF. This follows from the fact that the map $\,.^{\sim n}$ is $\mathbb{Q}$-linear and injective. □

DEFINITION 2.77. For $n \in \mathbb{N}$ and $\kappa \in \mathbb{K}_n$, we let $m_\kappa^{(n)} : \mathbb{K}_n \longrightarrow \mathbb{K}_n$ be the map that is given by multiplication by κ, i.e., $z \longmapsto \kappa z$.

REMARK 2.78. Let $n \in \mathbb{N}$. Note that $m_\kappa^{(n)}$ is a $\mathbb{Q}$-linear endomorphism of the $\phi(n)$-dimensional vector space $\mathbb{K}_n$ over $\mathbb{Q}$; cf. Proposition 1.11. Further, $m_\kappa^{(n)}$ is a $\mathbb{Q}$-linear automorphism if and only if $\kappa \neq 0$. In particular, if $n \in \mathbb{N} \setminus \{1,2,3,4,6\}$, then $m^{(n)}_{(\zeta_n+\bar{\zeta}_n)}$ is a $\mathbb{Q}$-linear automorphism, since then $\zeta_n + \bar{\zeta}_n \neq 0$. In fact, this restriction means that $\phi(n)/2 \geq 2$ from which follows that $\zeta_n + \bar{\zeta}_n \notin \mathbb{Q}$; cf. Corollary 1.14.

The $\mathbb{Q}$-linear endomorphism $m_\kappa^{(n)}$ (resp., automorphism, if $\kappa \neq 0$) corresponds via the $\mathbb{Q}$-linear isomorphism $^{\frown n}$ to a $\mathbb{Q}$-linear endomorphism (resp., automorphism), say $(m_\kappa^{(n)})^{\frown n}$, of $[\mathbb{K}_n]^{\sim n}$, i.e.,

$$(m_\kappa^{(n)})^{\frown n} = \,.^{\frown n} \circ m_\kappa^{(n)} \circ (\,.^{\frown n})^{-1}.$$

Note further that $(m_\kappa^{(n)})^{\frown n}$ extends uniquely to an $\mathbb{R}$-linear endomorphism (resp., automorphism), say $(m_\kappa^{(n)})^{\sim n}$, of $(\mathbb{R}^2)^{\phi(n)/2}$; cf. Remark 2.74.

The following result characterizes, for $n \in \mathbb{N} \setminus \{1,2\}$, the intersections of 1-dimensional $\mathcal{O}_n$-subspaces in the Euclidean plane with the nth cyclotomic field $\mathbb{K}_n$ in terms of existence of certain $\mathbb{Q}$-bases.

LEMMA 2.79. *Let $n \in \mathbb{N} \setminus \{1,2\}$ and let $o \in \mathcal{O}_n \setminus \{0\}$. Then, one has*

$$\mathbb{K}_n \cap (\mathbb{R}o) = \mathbb{k}_n o.$$

Moreover, $\mathbb{K}_n \cap (\mathbb{R}o)$ is a $(\phi(n)/2)$-dimensional $\mathbb{Q}$-linear subspace of $\mathbb{K}_n$ with $\mathbb{Q}$-basis $B_1^{(n)}o$.

PROOF. If $z \in \mathbb{K}_n \cap (\mathbb{R}o)$, one has $z = \lambda o$ with $\lambda \in \mathbb{k}_n$ by Proposition 1.23. This implies the equality $\mathbb{K}_n \cap (\mathbb{R}o) = \mathbb{k}_n o$. The additional statement now follows immediately from Corollary 1.14 together with our assumption $o \neq 0$. □

The next result will not be needed in the following. For clarity, we include it together with a proof.

PROPOSITION 2.80. *Let $n \in \mathbb{N} \setminus \{1,2\}$ and let $o \in \mathcal{O}_n \setminus \{0\}$. Then, one has:*

(a) *$\langle [\mathcal{O}_n \cap (\mathbb{R}o)]^{\sim n} \rangle_{\mathbb{R}}$ is a $(\phi(n)/2)$-dimensional $[\mathcal{O}_n]^{\sim n}$-subspace of $(\mathbb{R}^2)^{\phi(n)/2}$ with $\mathbb{R}$-basis*

$$[B_1^{(n)} o]^{\sim n} = \left\{ o^{\sim n}, (m^{(n)}_{(\zeta_n+\bar{\zeta}_n)})^{\frown n}(o^{\sim n}), \ldots, ((m^{(n)}_{(\zeta_n+\bar{\zeta}_n)})^{\frown n})^{\frac{\phi(n)}{2}-1}(o^{\sim n}) \right\}.$$

(b) *$\langle [\mathcal{O}_n \cap (\mathbb{R}o)]^{\sim n} \rangle_{\mathbb{R}} \cap [\mathbb{K}_n]^{\sim n} = [\Bbbk_n o]^{\sim n}$ and $[B_1^{(n)} o]^{\sim n}$ is a $\mathbb{Q}$-basis of the vector space*

$$\langle [\mathcal{O}_n \cap (\mathbb{R}o)]^{\sim n} \rangle_{\mathbb{R}} \cap [\mathbb{K}_n]^{\sim n}$$

over $\mathbb{Q}$.

PROOF. First, observe that

$$\begin{aligned} [B_1^{(n)} o]^{\sim n} &= \left\{ (b_1^{(n)} o)^{\sim n}, (b_2^{(n)} o)^{\sim n}, \ldots, (b^{(n)}_{\frac{\phi(n)}{2}} o)^{\sim n} \right\} \\ &= \left\{ o^{\sim n}, (m^{(n)}_{(\zeta_n+\bar{\zeta}_n)})^{\frown n}(o^{\sim n}), \ldots, ((m^{(n)}_{(\zeta_n+\bar{\zeta}_n)})^{\frown n})^{\frac{\phi(n)}{2}-1}(o^{\sim n}) \right\} \end{aligned}$$

and, trivially,

$$[B_1^{(n)} o]^{\sim n} \subset [\mathcal{O}_n \cap (\mathbb{R}o)]^{\sim n}.$$

Secondly, one can easily see that, since the elements $(b_j^{(n)})^{\sim n} \in (\mathbb{R}^2)^{\phi(n)/2}$, $j \in \{1, \ldots, \phi(n)/2\}$, are $\mathbb{R}$-linearly independent, the elements $(b_j^{(n)} o)^{\sim n} \in (\mathbb{R}^2)^{\phi(n)/2}$, $j \in \{1, \ldots, \phi(n)/2\}$, are $\mathbb{R}$-linearly independent as well. This shows that one has

$$\dim\left(\langle [\mathcal{O}_n \cap (\mathbb{R}o)]^{\sim n} \rangle_{\mathbb{R}} \right) \geq \tfrac{\phi(n)}{2}.$$

Assume that one has $\dim(\langle [\mathcal{O}_n \cap (\mathbb{R}o)]^{\sim n} \rangle_{\mathbb{R}}) > \phi(n)/2$, i.e., assume the existence of more than $\phi(n)/2$ elements of $[\mathcal{O}_n \cap (\mathbb{R}o)]^{\sim n}$ that are $\mathbb{R}$-linearly independent. The inclusion

$$[\mathcal{O}_n \cap (\mathbb{R}o)]^{\sim n} \subset [\mathbb{K}_n \cap (\mathbb{R}o)]^{\sim n},$$

then implies the existence of more than $\phi(n)/2$ elements of $[\mathbb{K}_n \cap (\mathbb{R}o)]^{\sim n}$ that are linearly independent over $\mathbb{R}$ and hence, linearly independent over $\mathbb{Q}$. This is a contradiction since, by Lemma 2.76 and Lemma 2.79, one has $\dim_{\mathbb{Q}}([\mathbb{K}_n \cap (\mathbb{R}o)]^{\sim n}) = \phi(n)/2$. Part (a) follows.

For part (b), we are done if we can show that the inclusion

$$\begin{aligned} [\Bbbk_n o]^{\sim n} = [\mathbb{K}_n \cap (\mathbb{R}o)]^{\sim n} &\subset \left\langle [\mathcal{O}_n \cap (\mathbb{R}o)]^{\sim n} \right\rangle_{\mathbb{R}} \cap [\mathbb{K}_n]^{\sim n} \\ &\subset \left\langle [\mathbb{K}_n \cap (\mathbb{R}o)]^{\sim n} \right\rangle_{\mathbb{R}} \cap [\mathbb{K}_n]^{\sim n} \\ &= \langle [\Bbbk_n o]^{\sim n} \rangle_{\mathbb{R}} \cap [\mathbb{K}_n]^{\sim n} \end{aligned}$$

holds and, moreover, that the left-hand side and the right-hand side of this inclusion are equal. Since, by Lemma 2.79, $\mathbb{K}_n \cap (\mathbb{R}o) = \Bbbk_n o$, the only non-trivial part of the inclusion is

$$[\mathbb{K}_n \cap (\mathbb{R}o)]^{\sim n} \subset \left\langle [\mathcal{O}_n \cap (\mathbb{R}o)]^{\sim n} \right\rangle_{\mathbb{R}} \cap [\mathbb{K}_n]^{\sim n}.$$

Let $z \in \mathbb{K}_n \cap (\mathbb{R}o)$. It follows that $z = \lambda o$ with $\lambda \in \Bbbk_n$ by Proposition 1.23. We consider the $\mathbb{Q}$-coordinates of λ with respect to the $\mathbb{Q}$-basis

$$\{1, (\zeta_n + \bar{\zeta}_n), (\zeta_n + \bar{\zeta}_n)^2, \ldots, (\zeta_n + \bar{\zeta}_n)^{\frac{\phi(n)}{2}-1}\}$$

of $\Bbbk_n$ (cf. Corollary 1.14) and let $k \in \mathbb{N}$ be the least common multiple of all their denominators. Then, by Remark 1.24, we get $k\lambda \in \mathcal{O}_n$ and $k\lambda o \in \mathcal{O}_n \cap (\mathbb{R}o)$. Since

$$(k\lambda o)^{\sim_n} = k(\lambda o)^{\sim_n} \in \left[\mathcal{O}_n \cap (\mathbb{R}o)\right]^{\sim_n},$$

one gets $z^{\sim_n} = (\lambda o)^{\sim_n} \in \langle [\mathcal{O}_n \cap (\mathbb{R}o)]^{\sim_n} \rangle_{\mathbb{R}}$. This proves the inclusion. In order to prove equality of the left-hand side and the right-hand side of this inclusion, consider the $\mathbb{Q}$-linear automorphism $(m^{(n)}_{1/o})^{\sim_n}$, of $[\mathbb{K}_n]^{\sim_n}$ together with its unique extension to an $\mathbb{R}$-linear automorphism, say $(m^{(n)}_{1/o})^{\sim_n}$, of $(\mathbb{R}^2)^{\phi(n)/2}$; cf. Remark 2.78. For the left-hand side, one has

$$(m^{(n)}_{\frac{1}{o}})^{\sim_n}\left[[\Bbbk_n o]^{\sim_n}\right] = [\Bbbk_n]^{\sim_n}.$$

Furthermore, for the right-hand side, one has

$$\begin{aligned}
(m^{(n)}_{\frac{1}{o}})^{\sim_n}\left[\langle [\Bbbk_n o]^{\sim_n} \rangle_{\mathbb{R}} \cap [\mathbb{K}_n]^{\sim_n}\right] &= (m^{(n)}_{\frac{1}{o}})^{\sim_n}\left[\langle [\Bbbk_n o]^{\sim_n} \rangle_{\mathbb{R}}\right] \cap (m^{(n)}_{\frac{1}{o}})^{\sim_n}\left[[\mathbb{K}_n]^{\sim_n}\right] \\
&= \left\langle (m^{(n)}_{\frac{1}{o}})^{\sim_n}\left[[\Bbbk_n o]^{\sim_n}\right] \right\rangle_{\mathbb{R}} \cap [\mathbb{K}_n]^{\sim_n} \\
&= \langle [\Bbbk_n]^{\sim_n} \rangle_{\mathbb{R}} \cap [\mathbb{K}_n]^{\sim_n}.
\end{aligned}$$

It remains to prove that

$$\langle [\Bbbk_n]^{\sim_n} \rangle_{\mathbb{R}} \cap [\mathbb{K}_n]^{\sim_n} = [\Bbbk_n]^{\sim_n}.$$

Clearly, one has $\langle [\Bbbk_n]^{\sim_n} \rangle_{\mathbb{R}} \cap [\mathbb{K}_n]^{\sim_n} \supset [\Bbbk_n]^{\sim_n}$, so we are only concerned with the other inclusion. Let $x \in \langle [\Bbbk_n]^{\sim_n} \rangle_{\mathbb{R}} \cap [\mathbb{K}_n]^{\sim_n}$. Since

$$\langle [\Bbbk_n]^{\sim_n} \rangle_{\mathbb{R}} = \left\langle \langle [B^{(n)}_1]^{\sim_n} \rangle_{\mathbb{Q}} \right\rangle_{\mathbb{R}} = \langle [B^{(n)}_1]^{\sim_n} \rangle_{\mathbb{R}}$$

and $[\mathbb{K}_n]^{\sim_n} = \langle [B^{(n)}]^{\sim_n} \rangle_{\mathbb{Q}}$, there are $\mu_1, \dots, \mu_{\phi(n)/2} \in \mathbb{R}$ and $q_1, \dots, q_{\phi(n)} \in \mathbb{Q}$ such that

$$x = \sum_{j=1}^{\frac{\phi(n)}{2}} \mu_j (b^{(n)}_j)^{\sim_n} = \sum_{j=1}^{\phi(n)} q_j (b^{(n)}_j)^{\sim_n}.$$

Since $[B^{(n)}]^{\sim_n}$ is an $\mathbb{R}$-basis of $(\mathbb{R}^2)^{\phi(n)/2}$, one gets $\mu_j = q_j$ and $q_{\phi(n)/2+j} = 0$ for all $j \in \{1, \dots, \phi(n)/2\}$. It follows that $x \in \langle [B^{(n)}_1]^{\sim_n} \rangle_{\mathbb{Q}} = [\Bbbk_n]^{\sim_n}$. This completes the proof. □

From now on, we consider $(\mathbb{R}^2)^{\phi(n)/2}$ as an Euclidean vector space, equipped with the inner product $\langle \cdot, \cdot \rangle$, defined by taking the basis $[B^{(n)}]^{\sim_n}$ of the lattice $[\mathcal{O}_n]^{\sim_n}$ to be an orthonormal basis. In particular, the orthogonal complement $T^{\perp}$ of a linear subspace T of $(\mathbb{R}^2)^{\phi(n)/2}$ and, further, the orthogonal projection $S|T$ of a subset S of $(\mathbb{R}^2)^{\phi(n)/2}$ on T are understood with respect to this inner product. Further, the norm $\|x\|$ of an element $x \in (\mathbb{R}^2)^{\phi(n)/2}$ is given by

$$\|x\| = \sqrt{\sum_{j=1}^{\phi(n)} \lambda_j^2}, \tag{2.6}$$

where the λ_j are the coordinates of x with respect to the basis $[B^{(n)}]^{\sim_n}$. We wish to emphasize that the latter should not be confused with the use of orthogonal projections in the notion of determination of finite sets by projections, where we always use the *canonical* inner product on $\mathbb{R}^d$, $d \geq 2$.

REMARK 2.81. Note that if $o \in \mathcal{O}_n$, then, by Proposition 1.23(a), one also has $\bar{o} \in \mathcal{O}_n$, where $\bar{o}$ denotes the complex conjugate of o.

LEMMA 2.82. *Let $n \in \mathbb{N} \setminus \{1,2\}$, let $F \in \mathcal{F}(\mathcal{O}_n)$ and let $u \in \mathbb{S}^1$ be an $\mathcal{O}_n$-direction, say parallel to $o \in \mathcal{O}_n \setminus \{0\}$. Then, the set*

$$G^F_{\{u\}} \cap \mathcal{O}_n$$

corresponds via the $\mathbb{Q}$-linear isomorphism

$$.^{\frown_n} \circ m^{(n)}_{\bar{o}} : \mathbb{K}_n \longrightarrow [\mathbb{K}_n]^{\frown_n}$$

to a subset of the lattice $[\mathcal{O}_n]^{\frown_n}$ which is contained in a finite union of translates of the form $y + [\mathcal{o}_n]^{\frown_n}$, where $y \in [\mathcal{O}_n]^{\frown_n}$.

PROOF. First, note that $m_{\bar{o}}$ is indeed a $\mathbb{Q}$-linear automorphism, since $\bar{o} \neq 0$; see Remark 2.78. Let $\ell \in \operatorname{supp}(X_u F)$ and consider an element $f + \lambda o \in \ell \cap \mathcal{O}_n$, where $f \in F \subset \mathcal{O}_n$ and $\lambda \in \mathbb{R}$. One sees that $\lambda o \in \mathcal{O}_n$ and, further, $\lambda \in \mathbb{k}_n$; recall that $\mathbb{k}_n$ is the maximal real subfield of $\mathbb{K}_n$. Since $\bar{o} \in \mathcal{O}_n$, one has $(\lambda o)\bar{o} \in \mathcal{o}_n$; cf. Proposition 1.23. This shows that $m^{(n)}_{\bar{o}}(f + \lambda o) = f\bar{o} + (\lambda o)\bar{o} \in f\bar{o} + \mathcal{o}_n$. Hence, one has $m^{(n)}_{\bar{o}}[\ell \cap \mathcal{O}_n] \in f\bar{o} + \mathcal{o}_n$. Consequently, $m^{(n)}_{\bar{o}}[G^F_{\{u\}} \cap \mathcal{O}_n]$ is contained in a finite union of translates of the form $a + \mathcal{o}_n$, where $a \in \mathcal{O}_n$. The assertion follows immediately. $\square$

REMARK 2.83. Note that $\langle [\mathcal{o}_n]^{\frown_n} \rangle_{\mathbb{R}}$ and $(\langle [\mathcal{o}_n]^{\frown_n} \rangle_{\mathbb{R}})^{\perp}$ are $(\phi(n)/2)$-dimensional $[\mathcal{O}_n]^{\frown_n}$-subspaces, since one has $\langle [\mathcal{o}_n]^{\frown_n} \rangle_{\mathbb{R}} = \langle [B^{(n)}_1]^{\frown_n} \rangle_{\mathbb{R}}$ and $(\langle [\mathcal{o}_n]^{\frown_n} \rangle_{\mathbb{R}})^{\perp} = \langle [B^{(n)}_2]^{\frown_n} \rangle_{\mathbb{R}}$.

In the next lemma, $(\langle [\mathcal{o}_n]^{\frown_n} \rangle_{\mathbb{R}})^{\perp}$ is seen as a metric space in the canonical way.

LEMMA 2.84. *Let $n \in \mathbb{N} \setminus \{1,2\}$. Furthermore, let B be an open ball of positive radius r in the $(\phi(n)/2)$-dimensional $[\mathcal{O}_n]^{\frown_n}$-subspace $(\langle [\mathcal{o}_n]^{\frown_n} \rangle_{\mathbb{R}})^{\perp}$. Then, there is a $(\phi(n)/2)$-dimensional $[\mathcal{O}_n]^{\frown_n}$-subspace S of $(\mathbb{R}^2)^{\phi(n)/2}$ with the following properties:*

(i) *For each $y \in [\mathcal{O}_n]^{\frown_n}$, the translate $y + S$ meets at most one point of*

$$[\mathcal{O}_n]^{\frown_n} \cap \left(B + \langle [\mathcal{o}_n]^{\frown_n} \rangle_{\mathbb{R}}\right).$$

(ii) *$S \cap [\mathbb{K}_n]^{\frown_n}$ has a $\mathbb{Q}$-basis of the form*

$$\left\{x, (m^{(n)}_{(\zeta_n + \bar{\zeta}_n)})^{\frown_n}(x), \dots, ((m^{(n)}_{(\zeta_n + \bar{\zeta}_n)})^{\frown_n})^{\frac{\phi(n)}{2}-1}(x)\right\},$$

where $x \in [\mathcal{O}_n]^{\frown_n} \setminus \{0\}$.

PROOF. For $\varepsilon \in \mathbb{Q}$ positive, set

$$S_\varepsilon := \left\{ x \in (\mathbb{R}^2)^{\frac{\phi(n)}{2}} \,\middle|\, \left\langle (b^{(n)}_j)^{\frown_n} + \varepsilon (b^{(n)}_{\frac{\phi(n)}{2}+j})^{\frown_n}, x \right\rangle = 0 \;\; \forall j \in \{1, \dots, \tfrac{\phi(n)}{2}\} \right\}.$$

Then, S_ε is a $[\mathcal{O}_n]^{\frown_n}$-subspace of dimension $\phi(n)/2$. Further, $[\mathbb{K}_n]^{\frown_n} \cap S_\varepsilon$ is a vector space of dimension $\phi(n)/2$ over $\mathbb{Q}$. Moreover, an $\mathbb{R}$-basis of S_ε which simultaneously is a $\mathbb{Q}$-basis of $[\mathbb{K}_n]^{\frown_n} \cap S_\varepsilon$ is given by

$$\begin{aligned} &\left\{ -\varepsilon (b^{(n)}_j)^{\frown_n} + (b^{(n)}_{\frac{\phi(n)}{2}+j})^{\frown_n} \,\middle|\, j \in \{1, \dots, \tfrac{\phi(n)}{2}\} \right\} \\ = \; &\left\{ ((m^{(n)}_{(\zeta_n + \bar{\zeta}_n)})^{\frown_n})^j \left(-\varepsilon (b^{(n)}_1)^{\frown_n} + (b^{(n)}_{\frac{\phi(n)}{2}+1})^{\frown_n} \right) \,\middle|\, j \in \{0, \dots, \tfrac{\phi(n)}{2} - 1\} \right\}. \end{aligned}$$

Note that there is even a suitable $x_\varepsilon \in [\mathcal{O}_n]^{\frown n} \setminus \{0\}$ such that

$$\left\{\left(m^{(n)}_{(\zeta_n+\bar{\zeta}_n)}\right)^{\frown n})^j(x_\varepsilon) \,|\, j \in \{0,\dots,\tfrac{\phi(n)}{2}-1\}\right\}$$

is an $\mathbb{R}$-basis of S_ε as well as a $\mathbb{Q}$-basis of $[\mathbb{K}_n]^{\frown n} \cap S_\varepsilon$. For example, if q_ε is the denominator of ε, then

$$x_\varepsilon := q_\varepsilon\Big(-\varepsilon(b_1^{(n)})^{\frown n} + \big(b^{(n)}_{\frac{\phi(n)}{2}+1}\big)^{\frown n}\Big)$$

has this property.

Suppose that $v \in ([\mathcal{O}_n]^{\frown n} \setminus \{0\}) \cap S_\varepsilon$, say $v = \sum_{j=1}^{\phi(n)} \lambda_j (b_j^{(n)})^{\frown n}$ with uniquely determined $\lambda_j \in \mathbb{Z}$; see Remark 2.74. Hence, one has

$$\lambda_j = -\varepsilon \lambda_{\frac{\phi(n)}{2}+j}$$

for $j \in \{1,\dots,\phi(n)/2\}$. Further, Since $v \neq 0$, one has $\lambda_j \neq 0$ for at least one $j \in \{1,\dots,\phi(n)\}$. If $v \notin (\langle [\mathcal{O}_n]^{\frown n}\rangle_\mathbb{R})^\perp$, there is a $j_0 \in \{1,\dots,\phi(n)/2\}$ such that $\lambda_{j_0} \neq 0$. It follows that

$$1 \le |\lambda_{j_0}| = \varepsilon\big|\lambda_{\frac{\phi(n)}{2}+j_0}\big| \le \varepsilon\|v\|.$$

If $v \in (\langle [\mathcal{O}_n]^{\frown n}\rangle_\mathbb{R})^\perp$, then $\lambda_j = 0$ for $j \in \{1,\dots,\phi(n)/2\}$. This means that there is a $j_0 \in \{1,\dots,\phi(n)/2\}$ with $\lambda_{\frac{\phi(n)}{2}+j_0} \neq 0$. Then, one has

$$0 = -\varepsilon\lambda_{\frac{\phi(n)}{2}+j_0} \neq 0\,,$$

a contradiction.

One sees that every $v \in ([\mathcal{O}_n]^{\frown n} \setminus \{0\}) \cap S_\varepsilon$ satisfies $\|v\| \ge 1/\varepsilon$. Further, one has

$$\Big\| v|\langle [\mathcal{O}_n]^{\frown n}\rangle_\mathbb{R} \Big\| = \varepsilon\sqrt{\Big(\lambda_{\frac{\phi(n)}{2}+1}\Big)^2 + \cdots + \Big(\lambda_{\phi(n)}\Big)^2} \le \varepsilon\|v\|\,.$$

It follows that, if $\varepsilon \le 1/2$, then

$$\Big\| v|(\langle [\mathcal{O}_n]^{\frown n}\rangle_\mathbb{R})^\perp \Big\| \ge \|v\| - \Big\| v|\langle [\mathcal{O}_n]^{\frown n}\rangle_\mathbb{R} \Big\| \ge \|v\| - \varepsilon\|v\| = (1-\varepsilon)\|v\| \ge \frac{1}{2\varepsilon}\,.$$

This implies that, if $v_1, v_2 \in [\mathcal{O}_n]^{\frown n}$ are distinct lattice points in a translate $y + S_\varepsilon$ of S_ε, where $y \in [\mathcal{O}_n]^{\frown n}$, and if $\varepsilon \le 1/2$, then the distance

$$\Big\| v_1|(\langle [\mathcal{O}_n]^{\frown n}\rangle_\mathbb{R})^\perp - v_2|(\langle [\mathcal{O}_n]^{\frown n}\rangle_\mathbb{R})^\perp \Big\| = \Big\| (v_1 - v_2)|(\langle [\mathcal{O}_n]^{\frown n}\rangle_\mathbb{R})^\perp \Big\|$$

between $v_1|(\langle [\mathcal{O}_n]^{\frown n}\rangle_\mathbb{R})^\perp$ and $v_2|(\langle [\mathcal{O}_n]^{\frown n}\rangle_\mathbb{R})^\perp$ is at least $1/(2\varepsilon)$.

Choose an element $\varepsilon_0 \in \mathbb{Q}$ such that

$$0 < \varepsilon_0 \le \min\left\{\frac{1}{2}, \frac{1}{4r}\right\}.$$

Then, for each $y \in [\mathcal{O}_n]^{\frown n}$, the translate $y + S_{\varepsilon_0}$ of S_{ε_0} meets at most one point of

$$[\mathcal{O}_n]^{\frown n} \cap \Big(B + \langle [\mathcal{O}_n]^{\frown n}\rangle_\mathbb{R}\Big)\,.$$

To see this, assume the existence of an element $y \in [\mathcal{O}_n]^{\sim n}$ and assume the existence of two distinct points v_1, v_2 in $(y + S_{\varepsilon_0}) \cap ([\mathcal{O}_n]^{\sim n} \cap (B + \langle [\mathcal{O}_n]^{\sim n} \rangle_{\mathbb{R}}))$. Then, one has

$$\begin{aligned} 2r &> \| v_1 | (\langle [\mathcal{O}_n]^{\sim n} \rangle_{\mathbb{R}})^{\perp} - v_2 | (\langle [\mathcal{O}_n]^{\sim n} \rangle_{\mathbb{R}})^{\perp} \| \\ &= \| (v_1 - v_2) | (\langle [\mathcal{O}_n]^{\sim n} \rangle_{\mathbb{R}})^{\perp} \| \geq \frac{1}{2\varepsilon_0} \geq 2r , \end{aligned}$$

a contradiction. We also saw above that a $\mathbb{Q}$-basis of $[\mathbb{K}_n]^{\sim n} \cap S_{\varepsilon_0}$ is given by

$$\left\{ (m^{(n)}_{(\zeta_n + \bar{\zeta}_n)})^{\frown n})^j (x_{\varepsilon_0}) \,\middle|\, j \in \{0, \dots, \tfrac{\phi(n)}{2} - 1\} \right\} .$$

This completes the proof. □

2.4.2. Successive determination of finite subsets of rings of cyclotomic integers. Now we are able to prove the first result of this section on successive determination.

THEOREM 2.85. *Let $n \in \mathbb{N} \setminus \{1, 2\}$. Then, one has:*

(a) *The set $\mathcal{F}(\mathcal{O}_n)$ is successively determined by two X-rays in $\mathcal{O}_n$-directions.*

(b) *The set $\mathcal{F}(\mathcal{O}_n)$ is successively determined by two projections on orthogonal complements of 1-dimensional $\mathcal{O}_n$-subspaces.*

PROOF. Let us first prove part (a). Let $F \in \mathcal{F}(\mathcal{O}_n)$ and let $u \in \mathbb{S}^1$ be an $\mathcal{O}_n$-direction. Let $o \in \mathcal{O}_n \setminus \{0\}$ be parallel to u. Suppose that $F' \in \mathcal{F}(\mathcal{O}_n)$ satisfies $X_u F = X_u F'$. Then, by Lemma 1.120, one has

$$F, F' \subset G^F_{\{u\}} \cap \mathcal{O}_n .$$

By Lemma 2.82, the set $G^F_{\{u\}} \cap \mathcal{O}_n$ corresponds via the $\mathbb{Q}$-linear isomorphism

$$.^{\frown n} \circ m^{(n)}_{\bar{o}} : \mathbb{K}_n \longrightarrow [\mathbb{K}_n]^{\sim n}$$

to a subset of the lattice $[\mathcal{O}_n]^{\sim n}$ which is contained in a finite union of translates of the form $y + [\mathcal{O}_n]^{\sim n}$, where $y \in [\mathcal{O}_n]^{\sim n}$. It follows that

$$(m^{(n)}_{\bar{o}})^{\frown n} [G^F_{\{u\}} \cap \mathcal{O}_n] | (\langle [\mathcal{O}_n]^{\sim n} \rangle_{\mathbb{R}})^{\perp}$$

is a finite set. Hence, it is contained in an open ball B of positive radius r in $(\langle [\mathcal{O}_n]^{\sim n} \rangle_{\mathbb{R}})^{\perp}$. By Lemma 2.84, there is a $(\phi(n)/2)$-dimensional $[\mathcal{O}_n]^{\sim n}$-subspace S of $(\mathbb{R}^2)^{\phi(n)/2}$ with the following properties:

(i) For each $y \in [\mathcal{O}_n]^{\sim n}$, the translate $y + S$ meets at most one point of

$$[\mathcal{O}_n]^{\sim n} \cap (B + \langle [\mathcal{O}_n]^{\sim n} \rangle_{\mathbb{R}}) .$$

(ii) $S \cap [\mathbb{K}_n]^{\sim n}$ has a $\mathbb{Q}$-basis of the form

$$\left\{ x, (m^{(n)}_{(\zeta_n + \bar{\zeta}_n)})^{\frown n}(x), \dots, ((m^{(n)}_{(\zeta_n + \bar{\zeta}_n)})^{\frown n})^{\frac{\phi(n)}{2} - 1}(x) \right\} ,$$

where $x \in [\mathcal{O}_n]^{\sim n} \setminus \{0\}$.

In particular, it follows that $S \cap [\mathbb{K}_n]^{\sim n}$ corresponds via $(.^{\frown n})^{-1}$ to a $\mathbb{Q}$-linear subspace of $\mathbb{K}_n$ of dimension $\phi(n)/2$, say L, with $\mathbb{Q}$-basis

$$\left\{ \left(m^{(n)}_{(\zeta_n + \bar{\zeta}_n)} \right)^j (o') \,\middle|\, j \in \{0, \dots, \tfrac{\phi(n)}{2} - 1\} \right\} ,$$

where o' is the unique element of $\mathcal{O}_n$ satisfying $(o')^{\sim_n} = x$. Clearly, one has $o' \in \mathcal{O}_n \setminus \{0\}$. Now, Lemma 2.79 immediately implies that

$$L = \mathbb{K}_n \cap (\mathbb{R}o').$$

Consider the element $(m_{\bar{o}}^{(n)})^{-1}(o') = o'/\bar{o} \in \mathbb{K}_n \setminus \{0\}$. By Proposition 1.11 and Remark 1.24, $(m_{\bar{o}}^{(n)})^{-1}(o')$ is parallel to a non-zero element of $\mathcal{O}_n$. Let $u' \in \mathbb{S}^1$ be an $\mathcal{O}_n$-direction parallel to $(m_{\bar{o}}^{(n)})^{-1}(o')$, e.g.,

$$u' := u_{(m_{\bar{o}}^{(n)})^{-1}(o')}.$$

We claim that $X_{u'}F = X_{u'}F'$ implies that $F = F'$. In order to prove our claim, we shall actually show that any line in the Euclidean plane of the form $\ell_{u'}^{o''}$ (cf. Definition 1.110), where $o'' \in \mathcal{O}_n$, meets at most one point of the set $G_{\{u\}}^F \cap \mathcal{O}_n$ defined above. To see this, assume the existence of an element $o'' \in \mathcal{O}_n$, and assume the existence of two distinct points g and g' in $\ell_{u'}^{o''} \cap (G_{\{u\}}^F \cap \mathcal{O}_n)$. We claim that $m_{\bar{o}}^{(n)}(g)$ and $m_{\bar{o}}^{(n)}(g')$ are two distinct points in

$$(o''\bar{o} + L) \cap m_{\bar{o}}^{(n)}\left[G_{\{u\}}^F \cap \mathcal{O}_n\right].$$

To see this, let $h \in \ell_{u'}^{o''} \cap (G_{\{u\}}^F \cap \mathcal{O}_n)$. It follows that there is a suitable $\lambda \in \mathbb{R}$ such that

$$h = o'' + \lambda\big(m_{\bar{o}}^{(n)}\big)^{-1}(o') \in G_{\{u\}}^F \cap \mathcal{O}_n \subset \mathcal{O}_n.$$

Proposition 1.23 implies that $\lambda \in \mathbb{k}_n$ and, moreover, one gets

$$m_{\bar{o}}^{(n)}(h) = m_{\bar{o}}^{(n)}\left(o'' + \lambda(m_{\bar{o}}^{(n)})^{-1}(o')\right) = o''\bar{o} + \lambda\frac{o'}{\bar{o}}\bar{o} = o''\bar{o} + \lambda o' \in o''\bar{o} + L.$$

This proves the claim. Finally, $(m_{\bar{o}}^{(n)}(g))^{\sim_n}$ and $(m_{\bar{o}}^{(n)}(g'))^{\sim_n}$ are two distinct points in

$$\begin{aligned}
&\left((o''\bar{o})^{\sim_n} + [L]^{\sim_n}\right) \cap \left[m_{\bar{o}}^{(n)}[G_{\{u\}}^F \cap \mathcal{O}_n]\right]^{\sim_n} \\
= \;&\left((o''\bar{o})^{\sim_n} + (S \cap [\mathbb{K}_n]^{\sim_n})\right) \cap \left[m_{\bar{o}}^{(n)}[G_{\{u\}}^F \cap \mathcal{O}_n]\right]^{\sim_n} \\
\subset \;&\left((o''\bar{o})^{\sim_n} + S\right) \cap \left[m_{\bar{o}}^{(n)}[G_{\{u\}}^F \cap \mathcal{O}_n]\right]^{\sim_n},
\end{aligned}$$

which contradicts property (i) above, since $(o'\bar{o})^{\sim_n} \in [\mathcal{O}_n]^{\sim_n}$ and since

$$\left[m_{\bar{o}}^{(n)}[G_{\{u\}}^F \cap \mathcal{O}_n]\right]^{\sim_n}$$

is a subset of $[\mathcal{O}_n]^{\sim_n} \cap (B + \langle [\mathcal{O}_n]^{\sim_n} \rangle_{\mathbb{R}})$. This proves part (a). Part (b) follows immediately from an analysis of the proof of part (a). □

COROLLARY 2.86. *Let $n \in \mathbb{N} \setminus \{1,2\}$. Then, one has:*

(a) *The set $\bigcup_{t \in \mathbb{R}^2} \mathcal{F}(t + \mathcal{O}_n)$ is successively determined by three X-rays in $\mathcal{O}_n$-directions.*
(b) *The set $\bigcup_{t \in \mathbb{R}^2} \mathcal{F}(t + \mathcal{O}_n)$ is successively determined by three projections on orthogonal complements of 1-dimensional $\mathcal{O}_n$-subspaces.*

PROOF. Let us first prove part (a). Let $U \subset \mathbb{S}^1$ be any set of two non-parallel $\mathcal{O}_n$-directions, say $u, u' \in \mathbb{S}^1$, having the property

(E) There are $o, o' \in \mathcal{O}_n \setminus \{0\}$ with $u_o = u$ and $u_{o'} = u'$, and satisfying one of the equivalent conditions (i)-(iii) of Proposition 1.129.

Clearly, there are sets U having property (E), e.g., $U := \{1, \zeta_n\}$ has this property. Let $F, F' \in \bigcup_{t\in\mathbb{R}^2} \mathcal{F}(t+\mathcal{O}_n)$, say $F \in \mathcal{F}(t+\mathcal{O}_n)$ and $F' \in \mathcal{F}(t'+\mathcal{O}_n)$, where $t, t' \in \mathbb{R}^2$, and suppose that F and F' have the same X-rays in the directions of U. Then, by Lemma 1.120 and Theorem 1.130 in conjunction with property (E), one obtains

$$F, F' \subset G_U^F \subset t + \mathcal{O}_n . \tag{2.7}$$

If $F = \varnothing$, then, by Lemma 1.112(a), one also gets $F' = \varnothing$. It follows that one may assume, without loss of generality, that F and F' are non-empty. Then, since $F' \subset t' + \mathcal{O}_n$, it follows from Equation (2.3) that $t + \mathcal{O}_n$ meets $t' + \mathcal{O}_n$, the latter being equivalent to the identity $t+\mathcal{O}_n = t'+\mathcal{O}_n$. Hence, one has $F - t, F' - t \in \mathcal{F}(\mathcal{O}_n)$, and, moreover, $F - t$ and $F' - t$ have the same X-rays in the directions of U. In particular, one has $X_u F = X_u F'$ for $u \in U$ from above. Now, beginning with this direction u, one obviously can proceed as in the proof of Theorem 2.85(a). Part (b) follows immediately from an analysis of the proof of part (a). □

2.4.3. Application to a class of Delone sets, cyclotomic model sets, and icosahedral model sets. Having the modelling of atomic constellations in mind, we note some immediate implications of Theorem 2.85 (resp., Corollary 2.86).

COROLLARY 2.87. *Let $n \in \mathbb{N}\setminus\{1,2\}$ and let $\varLambda \subset \mathbb{R}^2$ be a Delone set living on $\mathcal{O}_n$. Then, one has:*

(a) *The set $\mathcal{F}(\varLambda)$ (resp., $\bigcup_{t\in\mathbb{R}^2} \mathcal{F}(t+\varLambda)$) is successively determined by two (resp., three) X-rays in $\mathcal{O}_n$-directions.*
(b) *The set $\mathcal{F}(\varLambda)$ (resp., $\bigcup_{t\in\mathbb{R}^2} \mathcal{F}(t+\varLambda)$) is successively determined by two (resp., three) projections on orthogonal complements of 1-dimensional $\mathcal{O}_n$-subspaces.* □

COROLLARY 2.88. *Let $n \in \mathbb{N}\setminus\{1,2\}$ and let $\varLambda_n(t,W) \in \mathcal{M}(\mathcal{O}_n)$ be a cyclotomic model set. Then, one has:*

(a) *The set $\mathcal{F}(\varLambda_n(t,W))$ is successively determined by two X-rays in $\mathcal{O}_n$-directions.*
(b) *The set $\mathcal{F}(\varLambda_n(t,W))$ is successively determined by two projections on orthogonal complements of 1-dimensional $\mathcal{O}_n$-subspaces.* □

COROLLARY 2.89. *Let $n \in \mathbb{N}\setminus\{1,2\}$. Then, for all windows $W \subset (\mathbb{R}^2)^{\frac{\phi(n)}{2}-1}$, for all star maps $.^{\star_n} : \mathcal{O}_n \longrightarrow (\mathbb{R}^2)^{\phi(n)/2-1}$ (as described in Definition 1.73) and for all $R > 0$, one has:*

(a) *The set $\bigcup_{\varLambda\in\mathcal{M}_g^{\star_n}(W)} \mathcal{F}(\varLambda)$ is successively determined by three X-rays in $\mathcal{O}_n$-directions.*
(b) *The set $\bigcup_{\varLambda\in\mathcal{M}_g^{\star_n}(W)} \mathcal{F}(\varLambda)$ is successively determined by three projections on orthogonal complements of 1-dimensional $\mathcal{O}_n$-subspaces.* □

REMARK 2.90. Clearly, Theorem 2.85 and its immediate implications contained in Corollary 2.87 and Corollary 2.88 are optimal with respect to the number of X-rays (resp., projections) used. Note that $\mathcal{F}(\mathbb{R}^2)$ needs at least three X-rays (resp., projections) for its successive determination; see [**33**, Corollary 7.5]. Consequently, Corollary 2.87 and Corollary 2.88 show that cyclotomic model sets and even general Delone sets living on $\mathcal{O}_n$ are closer to lattices than to general point sets as far as successive determination is concerned; see [**33**, Corollary 7.3].

We are now able to give an alternative proof of Corollary 2.12.

COROLLARY 2.91 (cf. Corollary 2.12). *Let $n \in \mathbb{N} \setminus \{1,2\}$, let $\Lambda_n(t,W) \in \mathcal{M}(\mathcal{O}_n)$ be a cyclotomic model set, and let $R > 0$. Then, one has:*

(a) *The set $\mathcal{D}_{<R}(\Lambda_n(t,W))$ is determined by two X-rays in $\mathcal{O}_n$-directions.*
(b) *The set $\mathcal{D}_{<R}(\Lambda_n(t,W))$ is determined by two projections on orthogonal complements of 1-dimensional $\mathcal{O}_n$-subspaces.*

SECOND PROOF. We may assume, without loss of generality, that $t = 0$ and hence $\Lambda_n(t,W) \subset \mathcal{O}_n$. Let $.^{\sim n}$ be the Minkowski embedding of $\mathcal{O}_n$ that is used in the construction of $\Lambda_n(t,W)$, i.e., a map

$$.^{\sim n} : \mathcal{O}_n \longrightarrow \mathbb{R}^2 \times (\mathbb{R}^2)^{\frac{\phi(n)}{2}-1},$$

given by

$$z \longmapsto (z, z^{\star n}),$$

where

$$z^{\star n} = \left(\sigma_2(z), \dots, \sigma_{\frac{\phi(n)}{2}}(z)\right);$$

see Section 1.2.3.2. The assertion will follow from the ensuing analysis of the proof of Theorem 2.85. Let us first prove (a). Let $F \in \mathcal{D}_{<R}(\Lambda_n(t,W))$ and choose the $\mathcal{O}_n$-direction $u := 1 \in \mathbb{S}^1 \cap \mathcal{O}_n$. Suppose that $F' \in \mathcal{D}_{<R}(\Lambda_n(t,W))$ satisfies $X_1 F = X_1 F'$. Then, by Lemma 1.120, one has

$$F, F' \subset G^F_{\{1\}} \cap \Lambda_n(t,W) \subset \mathcal{O}_n.$$

An analysis of the proof of Lemma 2.82 shows that the set $G^F_{\{1\}} \cap \Lambda_n(t,W)$ maps via $.^{\sim n}$ to a subset of the lattice $[\mathcal{O}_n]^{\sim n}$ which is contained in a finite union of translates of the form $f^{\sim n} + [\mathcal{O}_n]^{\sim n}$, where $f \in F$. In particular, one has

$$\left[G^F_{\{1\}} \cap \Lambda_n(t,W)\right]^{\sim n} \subset [F]^{\sim n} + [\mathcal{O}_n]^{\sim n}$$

and, further,

$$\left[G^F_{\{1\}} \cap \Lambda_n(t,W)\right]^{\sim n} | (\langle [\mathcal{O}_n]^{\sim n} \rangle_{\mathbb{R}})^{\perp} \subset [F]^{\sim n} | (\langle [\mathcal{O}_n]^{\sim n} \rangle_{\mathbb{R}})^{\perp}. \tag{2.8}$$

Clearly, $[F]^{\sim n} | (\langle [\mathcal{O}_n]^{\sim n} \rangle_{\mathbb{R}})^{\perp}$ is a finite set. Further, by definition of $\Lambda_n(t,W)$, one has

$$[F]^{\sim n} = \{(f, f^{\star n}) \mid f \in F\} \subset F \times [F]^{\star n} \subset F \times W.$$

Since $F \in \mathcal{D}_{<R}(\Lambda_n(t,W))$ by assumption and since W is bounded (recall that $\mathrm{cl}(W)$ is compact), one sees that there is a positive $D \in \mathbb{R}$ such that, for all $F \in \mathcal{D}_{<R}(\Lambda_n(t,W))$, the set $[F]^{\sim n}$ has diameter at most D with respect to the maximum norm on $\mathbb{R}^2 \times (\mathbb{R}^2)^{\phi(n)/2-1}$, defined by the Euclidean norms on $\mathbb{R}^2$ and $(\mathbb{R}^2)^{\phi(n)/2-1} \cong \mathbb{R}^{\phi(n)-2}$, respectively. Since all norms on $\mathbb{R}^2 \times (\mathbb{R}^2)^{\phi(n)/2-1}$ are equivalent, there is a positive $D' \in \mathbb{R}$ such that, for all $F \in \mathcal{D}_{<R}(\Lambda_n(t,W))$, the set $[F]^{\sim n}$ has diameter at most D' with respect to the norm on $\mathbb{R}^2 \times (\mathbb{R}^2)^{\phi(n)/2-1}$ defined by the lattice $[\mathcal{O}_n]^{\sim n}$; cf. Equation (2.6). It follows from (2.8) that, for all $F \in \mathcal{D}_{<R}(\Lambda_n(t,W))$, the set $[G^F_{\{1\}} \cap \Lambda_n(t,W)]^{\sim n} | (\langle [\mathcal{O}_n]^{\sim n} \rangle_{\mathbb{R}})^{\perp}$ has diameter at most D' with respect to the induced norm on $(\langle [\mathcal{O}_n]^{\sim n} \rangle_{\mathbb{R}})^{\perp}$. Hence, there is a positive $r \in \mathbb{R}$ such that, for all $F \in \mathcal{D}_{<R}(\Lambda_n(t,W))$, the set $[G^F_{\{1\}} \cap \Lambda_n(t,W)]^{\sim n} | (\langle [\mathcal{O}_n]^{\sim n} \rangle_{\mathbb{R}})^{\perp}$ is contained in a suitable open ball B_F of radius r in $(\langle [\mathcal{O}_n]^{\sim n} \rangle_{\mathbb{R}})^{\perp}$. Observing that the $(\phi(n)/2)$-dimensional $[\mathcal{O}_n]^{\sim n}$-subspace S of $(\mathbb{R}^2)^{\phi(n)/2}$ in Lemma 2.84 does only depend on the radius r of the open

ball, one sees that it is possible here to choose the second direction u' independently from F and F'. This proves part (a). Part (b) again follows immediately from an analysis of the proof of part (a). □

REMARK 2.92. Similarly, one can modify [**33**, Lemma 7.1] in order to obtain an alternative proof of Corollary 2.14. Clearly, Corollary 2.91 is best possible with respect to the number of X-rays (resp., projections) used.

The following result is in full accordance with the setting described in Section 1.3.

THEOREM 2.93. *Let $n \in \mathbb{N} \setminus \{1,2\}$. Then, for all windows $W \subset (\mathbb{R}^2)^{\phi(n)/2-1}$, for all star maps $.^{\star_n}\colon \mathcal{O}_n \longrightarrow (\mathbb{R}^2)^{\phi(n)/2-1}$ (as described in Definition 1.73) and for all $R > 0$, one has:*

(a) *The set $\bigcup_{\varLambda \in \mathcal{M}_g^{\star_n}(W)} \mathcal{D}_{<R}(\varLambda)$ is determined by three X-rays in $\mathcal{O}_n$-directions.*
(b) *The set $\bigcup_{\varLambda \in \mathcal{M}_g^{\star_n}(W)} \mathcal{D}_{<R}(\varLambda)$ is determined by three projections on orthogonal complements of 1-dimensional $\mathcal{O}_n$-subspaces.*

PROOF. To prove part (a), let $U \subset \mathbb{S}^1$ be a set of two non-parallel $\mathcal{O}_n$-directions containing the $\mathcal{O}_n$-direction 1, say $U = \{1, u\}$. Suppose that $u \in \mathbb{S}^1$ has the property

(E') There is an element $o \in \mathcal{O}_n \setminus \{0\}$ with $u_o = u$ such that $\{1, o\}$ satisfy one of the equivalent conditions (i)-(iii) of Proposition 1.129.

Clearly, there are elements $u \in \mathbb{S}^1$ having property (E'), such as $u := \zeta_n \in \mathbb{S}^1$. Let

$$F, F' \in \bigcup_{\varLambda \in \mathcal{M}_g^{\star_n}(W)} \mathcal{D}_{<R}(\varLambda),$$

say $F \in \mathcal{D}_{<R}(\varLambda_n^{\star_n}(t, \tau + W))$ and $F' \in \mathcal{D}_{<R}(\varLambda_n^{\star_n}(t', \tau' + W))$, where $t, t' \in \mathbb{R}^2$ and $\tau, \tau' \in (\mathbb{R}^2)^{\phi(n)/2-1}$, and suppose that F and F' have the same X-rays in the directions of U. Then, by Lemma 1.120 and Theorem 1.130 in conjunction with property (E'), one obtains

$$F, F' \subset G_U^F \subset t + \mathcal{O}_n\,. \tag{2.9}$$

If $F = \varnothing$, then, by Lemma 1.112(a), one also gets $F' = \varnothing$. It follows that one may assume, without loss of generality, that F and F' are non-empty. Then, since $F' \subset t' + \mathcal{O}_n$, it follows from Equation (2.3) that $t + \mathcal{O}_n$ meets $t' + \mathcal{O}_n$, the latter being equivalent to the identity $t + \mathcal{O}_n = t' + \mathcal{O}_n$. Moreover, the identity $t + \mathcal{O}_n = t' + \mathcal{O}_n$ is equivalent to the relation $t' - t \in \mathcal{O}_n$. Hence, one has

$$F - t \in \mathcal{D}_{<R}\Big(\varLambda_n^{\star_n}(0, \tau + W)\Big)$$

and, since the equality $\varLambda_n^{\star_n}(t' - t, \tau' + W) = \varLambda_n^{\star_n}(0, (\tau' + (t' - t)^{\star_n}) + W)$ holds,

$$F' - t \in \mathcal{D}_{<R}\Big(\varLambda_n^{\star_n}(t' - t, \tau' + W)\Big) = \mathcal{D}_{<R}\Big(\varLambda_n^{\star_n}(0, (\tau' + (t' - t)^{\star_n}) + W)\Big)\,.$$

Clearly, $F - t$ and $F' - t$ again have the same X-rays in the directions of U. In particular, one has $X_1(F - t) = X_1(F' - t)$. Now, proceeding as in the proof of Corollary 2.91, one sees that there is an $\mathcal{O}_n$-direction u' such that the set $\bigcup_{\varLambda \in \mathcal{M}_g^{\star_n}(W)} \mathcal{D}_{<R}(\varLambda)$ is determined by the X-rays in the directions of the set $U' := U \cup \{u'\}$. In particular, the above set is determined by three X-rays in $\mathcal{O}_n$-directions. Part (b) follows immediately from an analysis of the proof of part (a). □

Applying Remark 1.98 in conjunction with Corollary 1.102, a careful analysis of the proof of Corollary 2.88 (resp., Corollary 2.89) immediately gives the following two implications:

COROLLARY 2.94. *Let* $\Lambda_{\text{ico}}(t, W)$ *be an icosahedral model set. Then, one has:*

(a) *The set* $\mathcal{F}(\Lambda_{\text{ico}}(t, W))$ *is successively determined by two* X*-rays in* $\mathrm{Im}[\mathbb{I}]^{(\tau,0,1)}$*-directions.*

(b) *The set* $\mathcal{F}(\Lambda_{\text{ico}}(t, W))$ *is successively determined by two projections on orthogonal complements of* 1*-dimensional* $\mathrm{Im}[\mathbb{I}]^{(\tau,0,1)}$*-subspaces.* □

REMARK 2.95. Clearly, Theorem 2.88 is optimal with respect to the number of X-rays (resp., projections) used.

COROLLARY 2.96. *For all windows* $W \subset \mathbb{R}^3$ *and for all* $R > 0$*, one has:*

(a) *The set* $\bigcup_{\Lambda \in \mathcal{I}_g(W)} \mathcal{F}(\Lambda)$ *is successively determined by three* X*-rays in* $\mathrm{Im}[\mathbb{I}]^{(\tau,0,1)}$*-directions.*

(b) *The set* $\bigcup_{\Lambda \in \mathcal{I}_g(W)} \mathcal{F}(\Lambda)$ *is successively determined by three projections on orthogonal complements of* 1*-dimensional* $\mathrm{Im}[\mathbb{I}]^{(\tau,0,1)}$*-subspaces.* □

CHAPTER 3

Complexity

The main algorithmic problems of discrete tomography of cyclotomic and icosahedral model sets Λ are discussed. In particular, it is shown that, when restricted to two Λ-directions, these problems are tractable for cyclotomic model sets and icosahedral model sets with polytopal windows.

3.1. The separation problem

The results of this section were found in cooperation with M. Baake, P. Gritzmann, B. Langfeld and K. Lord; see [**8**]. For convenience and completeness of the exposition, we repeat the details. First, the following convention should be born in mind.

Model of Computation. In the following, we apply the *real RAM-model* (Random-Access Machine) of computation, in which each storage location is capable of holding a single real number and where each of the following standard elementary operations on reals counts only with unit cost (unit time):

- The arithmetic operations ($+$, $-$, $\cdot$, $/$).
- Comparisons between two real numbers ($<$, $\leq$, $=$, $\neq$, $\geq$, $>$).
- Indirect addressing of memory (integer addresses only).
- $\sqrt[k]{\cdot}$, trigonometric functions, exp, and log (in general, analytic functions);

see [**54**] for details and further reading.

DEFINITION 3.1. Let $k \in \mathbb{N}$, let $P, W \subset \mathbb{R}^k$, and let $\tau \in \mathbb{R}^k$. We set

$$S_{W,\tau}(P) := P \cap (\tau + W)$$

and, further,

$$\mathrm{Sep}_W(P) := \left\{ S_{W,\tau}(P) \,\middle|\, \tau \in \mathbb{R}^k \right\} .$$

REMARK 3.2. Note that $\mathrm{Sep}_W(P)$ contains all subsets of P that are 'separable' from their complement (in P) by a translate of W. Trivially, one has $p \in \tau + W$ if and only if $\tau \in p - W$. It follows that

$$S_{W,\tau}(P) = \{ p \in P \,|\, \tau \in p - W \} ; \tag{3.1}$$

see Fig. 3.1 for an illustration.

When dealing with the consistency, reconstruction and uniqueness problems as defined in Definition 1.116, it is clear that, given a finite set P of points in H (the internal space) and a window $W \subset H$, we have to be able to decide efficiently whether P is contained in a translate of the interior $\mathrm{int}(W)$ of W. In the case of model sets with Euclidean internal space $H = \mathbb{R}^k$, $k \in \mathbb{N}$, this leads to the following geometric separation problem.

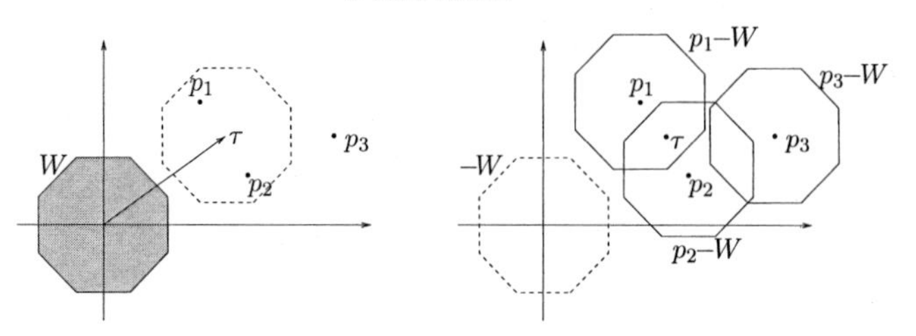

FIGURE 3.1. On the left: If we translate W by τ, then $\{p_1, p_2\}$ is a subset of $\tau + W$, but $\{p_3\}$ is not. On the right: The point τ is contained in $p_1 - W$ and $p_2 - W$, but not in $p_3 - W$. Again, we see that $S_{W,\tau}(P) = \{p_1, p_2\}$.

DEFINITION 3.3 (Separation Problem). Let $k \in \mathbb{N}$ and let $W \subset \mathbb{R}^k$. The corresponding separation problem is defined as follows.

SEPARATION.
Given a finite set $F \subset \mathbb{R}^k$, determine $\mathrm{Sep}_W(F)$.

We shall now show how to deal with the problem SEPARATION in case of windows W that are open polyhedra, i.e.,

$$W = \{x \in \mathbb{R}^k \mid Ax < b\} \quad \text{with } A \in \mathbb{R}^{l \times k} \text{ and } b \in \mathbb{R}^l ,$$

where $k \geq 2$ is a fixed constant. Note that the presented ideas can be generalized to semi-algebraic sets. Note further that all aperiodic cyclotomic model sets in Example 1.76 have polytopal windows.

3.1.1. Hyperplane arrangements. We summarize some facts about hyperplane arrangements as they are needed to deal with SEPARATION. See **[26]** for more information on hyperplane arrangements, and **[1, 40]** for surveys that cover also more general classes of arrangements.

DEFINITION 3.4. For $k, l \in \mathbb{N}$ and $j \in \{1, \ldots, l\}$, let $a_j \in \mathbb{R}^k \setminus \{0\}$, $\beta_j \in \mathbb{R}$, and consider the sets $H_j := \{x \in \mathbb{R}^k \mid a_j^t x = \beta_j\}$. Then, H_j is called a *hyperplane* and $\mathcal{H} := \{H_1, \ldots, H_l\}$ is called a *hyperplane arrangement* in $\mathbb{R}^k$. The *sign vector* $SV(x) \in \{\pm 1, 0\}^{\mathcal{H}}$ of a point $x \in \mathbb{R}^k$ is defined component wise via

$$SV_{\mathcal{H}_j}(x) := \begin{cases} -1 & \text{if } a_j^t x < \beta_j \\ 0 & \text{if } a_j^t x = \beta_j \\ +1 & \text{if } a_j^t x > \beta_j \end{cases} , \quad 1 \leq j \leq l .$$

If $s \in \{\pm 1, 0\}^{\mathcal{H}}$ satisfies $C_s := \{x \in \mathbb{R}^k \mid SV(x) = s\} \neq \varnothing$, then C_s is called a (proper) *cell* of the arrangement $\mathcal{H}$.

REMARK 3.5. The cells of a hyperplane arrangement $\mathcal{H}$ are relatively open sets of various dimensions. In particular, a cell C_s with sign vector s is full-dimensional if and only if $s \in \{\pm 1\}^{\mathcal{H}}$. Of course, $\mathbb{R}^k$ is the disjoint union of the cells of a hyperplane arrangement; see Figure 3.2 for an illustration.

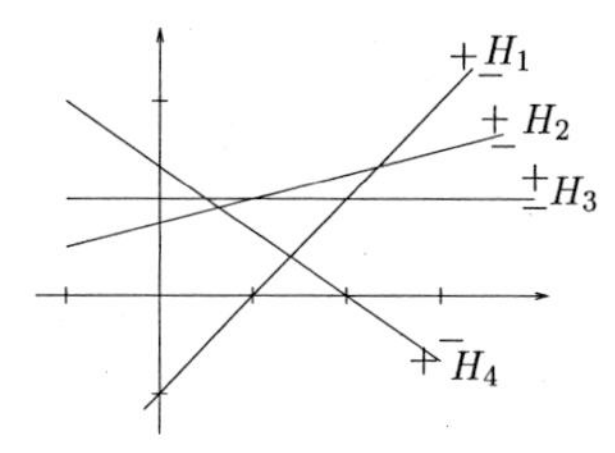

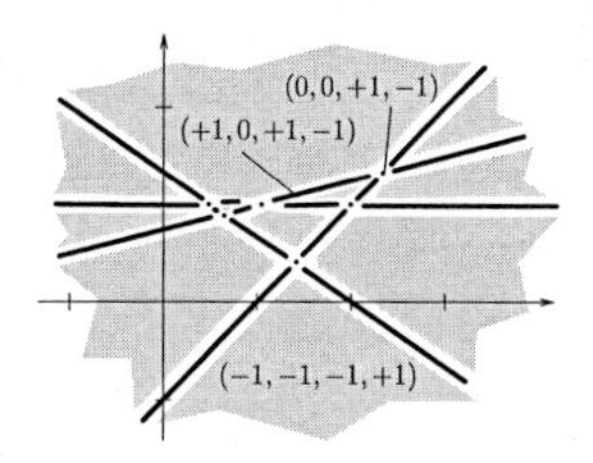

FIGURE 3.2. A hyperplane-arrangement in the plane. The hyperplanes are given by $H_j := \{x \in \mathbb{R}^2 \mid a_j^t x = \beta_j\}$, where $a_1^t = (-1,1)$, $\beta_1 = -1$, $a_2^t = (-1,4)$, $\beta_2 = 3$, $a_3^t = (0,1)$, $\beta_3 = 1$, $a_4^t = (-2,-3)$, $\beta_4 = -4$. On the right, the cells are drawn schematically. The arrangement consists of six points, 16 one-dimensional cells (thick lines) and 11 full-dimensional cells (grey areas). Some sign vectors are given. Note that not all vectors in $\{\pm 1, 0\}^4$ occur as sign vectors of cells; e.g., $(0,0,0,0)$ and $(-1,+1,-1,-1)$ are not realized.

In view of their general relevance, hyperplane arrangements are well studied and also algorithmically well understood. Let us recall the following result, which is based on [**26**, Thm. 3.3] and [**27**, Ch. 7].

PROPOSITION 3.6. *Let $k, l \in \mathbb{N}$ and let $\mathcal{H} = \{H_1, \ldots, H_l\}$ be a hyperplane arrangement in $\mathbb{R}^k$. There exists an algorithm that computes a set of points meeting each cell of $\mathcal{H}$ in $O(l^k)$ operations.* □

3.1.2. Tractability of the separation problem for polytopal windows. The following observation ties the separation problem to certain hyperplane arrangements.

LEMMA 3.7. *Let $k, l, q \in \mathbb{N}$, let $P = \{p_1, \ldots, p_q\}$ be a finite set of points in $\mathbb{R}^k$, let $W = \{x \in \mathbb{R}^k \mid Ax < b\}$ with $A \in \mathbb{R}^{l \times k}$, $b \in \mathbb{R}^l$, and let a_j^t denote the jth row of A, $1 \le j \le l$. For $1 \le j \le l$ and $1 \le r \le q$, set*

$$H_j^{(r)} := \{x \in \mathbb{R}^k \mid a_j^t x = (Ap_r - b)_j\}$$

Further, set

$$\mathcal{H}(W, P) := \{H_j^{(r)} \mid 1 \le j \le l \ , \ 1 \le r \le q\} \ .$$

Then, one has:

(a) *The set $p_r - W$ is an intersection of open halfspaces defined by the hyperplanes $H_1^{(r)}, \ldots, H_l^{(r)}$, more precisely, one has*

$$p_r - W = \{x \in \mathbb{R}^k \mid A^t x > Ap_r - b\} \ .$$

(b) *For all $s \in \{\pm 1, 0\}^{\mathcal{H}(W,P)}$ and all cells C_s of the hyperplane arrangement $\mathcal{H}(W, P)$ with sign vector s, the following implication is true:*

$$\tau, \tau' \in C_s \quad \Longrightarrow \quad S_{W,\tau}(P) = S_{W,\tau'}(P) \ .$$

(In general, the reverse implication is not true; see Figure 3.3.)

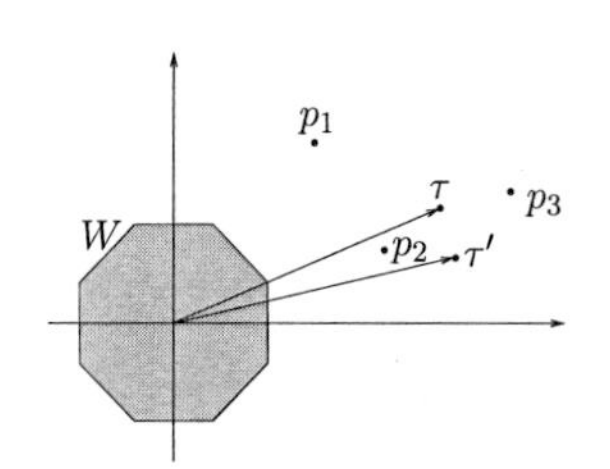

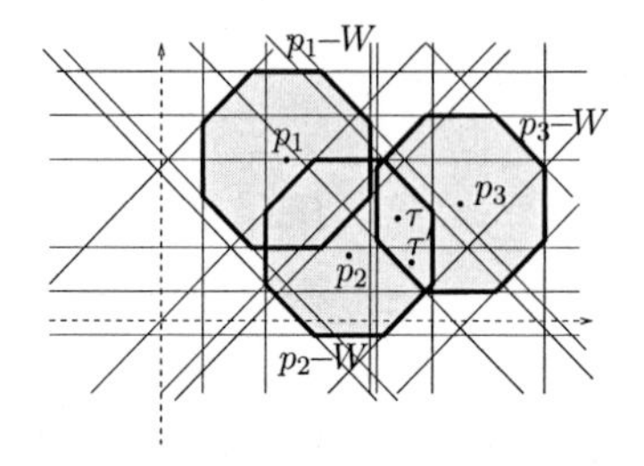

FIGURE 3.3. This shows in essence the same situation as in Fig. 3.1, on the right we added the arrangement $\mathcal{H}(W,P)$. Here, $S_{W,\tau}(P) = S_{W,\tau'}(P)$, but τ and τ' are in different fulldimensional cells of the arrangement $\mathcal{H}(W,P)$.

PROOF. Part (a) follows from a simple computation. For $p \in P$, one has

$$\begin{aligned} p - W &= \{p - x \mid x \in \mathbb{R}^k,\ Ax < b\} \\ &= \{x \in \mathbb{R}^k \mid A(p-x) < b\} \\ &= \{x \in \mathbb{R}^k \mid Ax > Ap - b\}. \end{aligned}$$

For (b), recall from Equation (3.1) that, for any $\tau \in \mathbb{R}^k$, one has

$$S_{W,\tau}(P) = \{p_r \in P \mid 1 \le r \le q\ ,\ \tau \in p_r - W\}.$$

Using (a) we see that $\tau \in p_r - W$ if and only if

$$SV_{H_1^{(r)}}(\tau) = \cdots = SV_{H_l^{(r)}}(\tau) = +1.$$

Now, if $\tau, \tau' \in C_s$, we have $SV(\tau) = SV(\tau') = s$. This completes the proof. □

THEOREM 3.8. *Let $k, l, q \in \mathbb{N}$, and let $W = \{x \in \mathbb{R}^k \mid Ax < b\}$ with $A \in \mathbb{R}^{l\times k}$ and $b \in \mathbb{R}^l$. Moreover, let $P = \{p_1, \ldots, p_q\}$ be a finite set of points in $\mathbb{R}^k$. Then,* $\mathrm{Sep}_W(P)$ *can be computed with at most $O((lq)^{k+1})$ operations.*

PROOF. One possible algorithm to determine $\mathrm{Sep}_W(P)$ performs the following steps.
Step 1: Compute $(Ap_r - b)_j$ for $1 \le j \le l$ and $1 \le r \le q$, to specify the hyperplane arrangement $\mathcal{H}(W,P)$ from Lemma 3.7.
Step 2: Compute a set T of points meeting every cell of $\mathcal{H}(W,P)$.
Step 3: For each of the points $\tau \in T$ obtained in Step 2, compute $S_{W,\tau}(P)$.
Step 4: Output the collection of all the $S_{W,\tau}(P)$.

The correctness of this procedure follows directly from Lemma 3.7.

Now we show the complexity assertion. Step 1 needs no more than $O(lq)$ operations. Step 2 requires $O((lq)^k)$ operations by Proposition 3.6. For Step 3, decide whether $\tau \in p_r - W$ for each r. To this end, test whether τ satisfies the inequalities $a_j^t\tau > (Ap_r - b)_j$, $1 \le j \le l$, $1 \le r \le q$. This can be done with $O(lq)$ operations. In total, this does not need more than $O(lq + (lq)^k lq) = O((lq)^{k+1})$ operations. □

REMARK 3.9. As the proof of Theorem 3.8 shows, if the number of hyperplanes defining the window W is regarded constant, then

$$\mathrm{card}\left(\mathrm{Sep}_W(P)\right) = O\left(\mathrm{card}(P)^k\right).$$

Theorem 3.8 can be generalized to semialgebraic sets W. The corresponding algorithm is then based on an analogue of Proposition 3.6; see [**16**] and [**17**, Theorem 2].

Theorem 3.8 immediately implies the following variant:

THEOREM 3.10. *Let $k, l \in \mathbb{N}$ and let $W = \{x \in \mathbb{R}^k \mid Ax \leq b\}$ with $A \in \mathbb{R}^{l \times k}$ and $b \in \mathbb{R}^l$. (The parameters k, l, A, and b are not part of the input). Then, for any finite set $P \subset \mathbb{R}^k$, the problem of computing $\mathrm{Sep}_{\mathrm{int}(W)}(P)$ can be solved in $O(\mathrm{card}(P)^{k+1})$ operations.* □

3.2. The decomposition problem

3.2.1. The general setting.

DEFINITION 3.11 (Decomposition Problem). Let $d \geq 2$, let $\varLambda \subset \mathbb{R}^d$ be a model set with underlying $\mathbb{Z}$-module L, and let $u_1, \dots, u_m \in \mathbb{S}^{d-1}$ be $m \geq 2$ pairwise non-parallel $\varLambda$-directions. The corresponding decomposition problem is defined as follows.

> DECOMPOSITION.
> Given functions $p_{u_j} : \mathcal{L}^d_{u_j} \longrightarrow \mathbb{N}_0$, $j \in \{1, \dots, m\}$, with finite supports that satisfy $\mathrm{supp}(p_{u_j}) \subset \mathcal{L}^L_{u_j}$, compute the equivalence classes modulo L in the associated grid $G_{\{p_{u_j} | j \in \{1,\dots,m\}\}}$.

3.2.2. Cyclotomic model sets. For cyclotomic model sets, the decomposition problem looks as follows; cf. Lemma 1.84.

DEFINITION 3.12 (Decomposition Problem). Let $n \in \mathbb{N} \setminus \{1, 2\}$ and let $u_1, \dots, u_m \in \mathbb{S}^1$ be $m \geq 2$ pairwise non-parallel $\mathcal{O}_n$-directions. The corresponding decomposition problem is defined as follows.

> DECOMPOSITION.
> Given functions $p_{u_j} : \mathcal{L}^2_{u_j} \longrightarrow \mathbb{N}_0$, $j \in \{1, \dots, m\}$, with finite supports that satisfy $\mathrm{supp}(p_{u_j}) \subset \mathcal{L}^{\mathcal{O}_n}_{u_j}$, compute the equivalence classes modulo $\mathcal{O}_n$ in the associated grid $G_{\{p_{u_j} | j \in \{1,\dots,m\}\}}$.

REMARK 3.13. The phenomenon of multiple equivalence classes modulo $\mathcal{O}_n$ in grids of the form G^F_U, where $F \subset \mathcal{O}_n$ and U is a finite set of pairwise non-parallel $\mathcal{O}_n$-directions, occurs already in the classical lattice situation; see Figure 1.6 on the left. There, *no* translate of the finite subset of the square lattice (marked by the connecting lines) is contained in any of the other equivalence classes. Hence, the problem of decomposing the grid into its equivalence classes modulo $\mathcal{O}_n$ is the first problem to be solved when dealing with the consistency or the reconstruction problem.

3.2.2.1. *Tractability of the decomposition problem for cyclotomic model sets.*

REMARK 3.14. Let $n \in \mathbb{N} \setminus \{1,2\}$ and let $u_1, \dots, u_m \in \mathbb{S}^1$ be $m \geq 2$ pairwise non-parallel $\mathcal{O}_n$-directions. Note that, for any instance of the corresponding decomposition problem, the associated grid $G_{\{p_{u_j} | j \in \{1,\dots,m\}\}}$ satisfies

$$\operatorname{card}\left(G_{\{p_{u_j} | j \in \{1,\dots,m\}\}}\right) \leq s^2 ,$$

where

$$s := \max\left(\{\operatorname{card}(\operatorname{supp}(p_{u_j})) \mid j \in \{1,\dots,m\}\}\right).$$

Since $G_{\{p_{u_j} | j \in \{1,\dots,m\}\}} \subset G^{\mathcal{O}_n}_{\{u_1,u_2\}}$, Proposition 1.122 in conjunction with Proposition 1.125 show that, for any set $\{u_1, \dots, u_m\}$ of $m \geq 2$ pairwise non-parallel $\mathcal{O}_n$-directions, there is a constant $c \in \mathbb{N}$ such that, for any instance of the corresponding decomposition problem, the associated grid $G_{\{p_{u_j} | j \in \{1,\dots,m\}\}}$ contains at most c equivalence classes modulo $\mathcal{O}_n$.

In the following, we assume that the elements of the supports of the functions p_{u_j}, $j \in \{1, \dots, m\}$, are given in the form $o + \mathbb{R}o_j$ for suitable $o \in \mathcal{O}_n$ and $o_j \in \mathcal{O}_n \setminus \{0\}$. Moreover, we assume that all o's and all o_j are given through their $\mathbb{Z}$-coordinates with respect to the $\mathbb{Z}$-basis $\{1, \zeta_n, \zeta_n^2, \dots, \zeta_n^{\phi(n)-1}\}$ of $\mathcal{O}_n$ (cf. Remark 1.24). The next algorithmic result shows that the decomposition problem for cyclotomic model sets can be solved efficiently.

THEOREM 3.15. *The decomposition problem can be solved in polynomial time. More precisely, it is of complexity $O(s^2)$, where s is the maximum of the cardinalities of the supports of the given functions p_{u_j}.*

PROOF. The algorithm performs the following steps.

Step 1: By the proof of Lemma 1.9(a), the Euclidean algorithm in $\mathbb{Z}[X]$, the inductive computability of the nth cyclotomic polynomial $F_n = \operatorname{Mipo}_{\mathbb{Q}}(\zeta_n)$ (cf. Remark 1.27 and Proposition 1.28), the proof of Proposition 1.122 and the Gaussian elimination algorithm, we are able to compute the $\mathbb{Q}$-coordinates of the elements of the grid $G_{\{p_{u_j} | j \in \{1,\dots,m\}\}} \subset \mathbb{K}_n$ (cf. Proposition 1.122) with respect to the $\mathbb{Q}$-basis $\{1, \zeta_n, \zeta_n^2, \dots, \zeta_n^{\phi(n)-1}\}$ of $\mathbb{K}_n$ (cf. Proposition 1.11) efficiently.

Step 2: Since $\{1, \zeta_n, \zeta_n^2, \dots, \zeta_n^{\phi(n)-1}\}$ is simultaneously a $\mathbb{Q}$-basis of $\mathbb{K}_n$ and a $\mathbb{Z}$-basis of $\mathcal{O}_n$ (cf. Proposition 1.11 and Remark 1.24), one has for all $q_0, q_1, \dots, q_{\phi(n)-1} \in \mathbb{Q}$ the equivalence

$$q_0 + q_1\zeta_n + \dots + q_{\phi(n)-1}\zeta_n^{\phi(n)-1} \in \mathcal{O}_n \iff q_0, q_1, \dots, q_{\phi(n)-1} \in \mathbb{Z}. \tag{3.2}$$

By Step 1, the elements of the grid $G_{\{p_{u_j} | j \in \{1,\dots,m\}\}}$ are given in the form

$$q_0 + q_1\zeta_n + \dots + q_{\phi(n)-1}\zeta_n^{\phi(n)-1},$$

where $q_0, q_1, \dots, q_{\phi(n)-1} \in \mathbb{Q}$. Now, proceed as follows: choose an arbitrary element g of the grid $G_{\{p_{u_j} | j \in \{1,\dots,m\}\}}$ and compute the $\mathbb{Q}$-coordinates of the differences $g - h$ with respect to $\{1, \zeta_n, \zeta_n^2, \dots, \zeta_n^{\phi(n)-1}\}$, where $h \in G_{\{p_{u_j} | j \in \{1,\dots,m\}\}} \setminus \{g\}$. By the above criterion (3.2), a fixed h lies in the same equivalence class modulo $\mathcal{O}_n$ as g if and only if all coordinates of $g - h$

are elements of $\mathbb{Z}$. Iterate this procedure by successively removing the computed equivalence classes and proceeding with the remaining subset of the grid and an arbitrary element therein.

By Remark 3.14, there is a constant $c \in \mathbb{N}$ such that, for all instances of the decomposition problem, Step 2 of this algorithm computes the equivalence classes of the grid modulo $\mathcal{O}_n$ in at most c iterations. The inequality

$$\operatorname{card}\left(G_{\{p_{u_j} \mid j \in \{1,\dots,m\}\}}\right) \le s^2$$

(cf. Remark 3.14 again) now completes the proof. □

REMARK 3.16. The proof of Theorem 3.15 indicates that we actually do not need the full strength of the real RAM-model of computation. Rather, a Turing machine model that is augmented for algebraic computations suffices; see [**20**]. Then, of course, the binary size of the input matters.

3.2.3. Icosahedral model sets. For icosahedral model sets, the decomposition problem looks as follows; cf. Lemma 1.109.

DEFINITION 3.17 (Decomposition Problem). Let $u_1, \dots, u_m \in \mathbb{S}^2$ be $m \ge 2$ pairwise non-parallel $\mathrm{Im}[\mathbb{I}]$-directions. The corresponding decomposition problem is defined as follows.

> DECOMPOSITION.
> Given functions $p_{u_j} : \mathcal{L}^3_{u_j} \longrightarrow \mathbb{N}_0$, $j \in \{1, \dots, m\}$, with finite supports that satisfy $\operatorname{supp}(p_{u_j}) \subset \mathcal{L}^{\mathrm{Im}[\mathbb{I}]}_{u_j}$, compute the equivalence classes modulo $\mathrm{Im}[\mathbb{I}]$ in the grid $G_{\{p_{u_j} \mid j \in \{1,\dots,m\}\}}$.

3.2.3.1. *Tractability of the decomposition problem for icosahedral model sets.*

REMARK 3.18. Let $u_1, \dots, u_m \in \mathbb{S}^2$ be $m \ge 2$ pairwise non-parallel $\mathrm{Im}[\mathbb{I}]^{(\tau,0,1)}$-directions. Note that, for any instance of the corresponding decomposition problem, the associated grid $G_{\{p_{u_j} \mid j \in \{1,\dots,m\}\}}$ satisfies

$$\operatorname{card}\left(G_{\{p_{u_j} \mid j \in \{1,\dots,m\}\}}\right) \le s^2 \,,$$

where

$$s := \max\left(\{\operatorname{card}(\operatorname{supp}(p_{u_j})) \mid j \in \{1, \dots, m\}\}\right).$$

Since $G_{\{p_{u_j} \mid j \in \{1,\dots,m\}\}} \subset G^{\mathrm{Im}[\mathbb{I}]}_{\{u_1,u_2\}}$, Proposition 1.131 in conjunction with Proposition 1.134 show that, for any set $\{u_1, \dots, u_m\}$ of $m \ge 2$ pairwise non-parallel $\mathrm{Im}[\mathbb{I}]^{(\tau,0,1)}$-directions, there is a constant $c \in \mathbb{N}$ such that, for any instance of the corresponding decomposition problem, the associated grid $G_{\{p_{u_j} \mid j \in \{1,\dots,m\}\}}$ contains at most c equivalence classes modulo $\mathrm{Im}[\mathbb{I}]$.

In the following, we assume that the elements of the supports of the functions p_{u_j}, $j \in \{1, \dots, m\}$, are always given in the form $\alpha + \mathbb{R}\alpha_{u_j}$ for suitable $\alpha \in \mathrm{Im}[\mathbb{I}]$ and $\alpha_{u_j} \in \mathrm{Im}[\mathbb{I}] \setminus \{0\}$. Moreover, we assume that all α's and all α_{u_j} are given through their $\mathbb{Z}$-coordinates with respect to the $\mathbb{Z}$-basis $B_{\mathrm{Im}[\mathbb{I}]}$ of $\mathrm{Im}[\mathbb{I}]$ (cf. Corollary 1.100).

THEOREM 3.19. *When restricted to* $\mathrm{Im}[\mathbb{I}]^{(\tau,0,1)}$*-directions, the decomposition problem can be solved in polynomial time. More precisely, it is of complexity* $O(s^2)$*, where* s *is the maximum of the cardinalities of the supports of the given functions* p_{u_j}.

PROOF. The algorithm performs the following steps.

Step 1: Compute the partition of the set $\bigcup_{j=1}^{m} \operatorname{supp}(p_{u_j})$ with respect to the equivalence relation $\equiv$ which is defined by

$$\ell \equiv \ell' \;:\Longleftrightarrow\; \ell - \ell' \subset H^{(\tau,0,1)}\,,$$

i.e., one has $\ell \equiv \ell'$ if and only if ℓ and ℓ' lie in a common translate of $H^{(\tau,0,1)}$ in $\mathbb{R}^3$. As a consequence, one obtains, say,

$$\bigcup_{j=1}^{m} \operatorname{supp}(p_{u_j}) = \dot{\bigcup}_{k=1}^{t} S_k$$

for a suitable $t \in \mathbb{N}$. This can be done efficiently.

Step 2: For all $k \in \{1,\dots,t\}$, compute the image $\Phi[\bigcup_{\ell\in S_k}(\ell-\alpha_k)]$ under the $\mathbb{R}$-linear map Φ (as defined in Definition 1.97(a)), where $\alpha_k \in \operatorname{Im}[\mathbb{I}]$ occurs as the supporting vector of some line $\ell \in S_k$. More concrete, for all $k \in \{1,\dots,t\}$ and all $\ell \in S_k$, compute the $\mathbb{Z}$-coordinates of suitable elements $z \in \mathcal{O}_5$ and $z_{u_j} \in \mathcal{O}_5 \setminus \{0\}$ with respect to the $\mathbb{Z}$-basis $\{1,\zeta_5,\zeta_5^2,\zeta_5^3\}$ of $\mathcal{O}_5$ such that

$$\Phi\left[\ell-\alpha_k\right] = z + \mathbb{R}z_{u_j}\,.$$

This can be done efficiently. Indeed, by assumption, every line ℓ in some S_k is given by $\ell = \alpha + \mathbb{R}\alpha_{u_j}$, where $\alpha \in \operatorname{Im}[\mathbb{I}]$ and $\alpha_{u_j} \in \operatorname{Im}[\mathbb{I}] \setminus \{0\}$ are given through their $\mathbb{Z}$-coordinates with respect to the $\mathbb{Z}$-basis $B_{\operatorname{Im}[\mathbb{I}]}$ of $\operatorname{Im}[\mathbb{I}]$. Since $\ell - \alpha_k = (\alpha-\alpha_k) + \mathbb{R}\alpha_{u_j}$, one only has to compute the $\mathbb{Z}$-coordinates of $\ell-\alpha_k$ with respect to $B_{\operatorname{Im}[\mathbb{I}]}$. The non-zero coordinates of $\ell-\alpha_k$ and α_{u_j} with respect to $B_{\operatorname{Im}[\mathbb{I}]}$ are, when suitably ordered, already the desired $\mathbb{Z}$-coordinates of $z := \Phi(\alpha-\alpha_k)$ and $z_{u_j} := \Phi(\alpha_{u_j})$ with respect to the $\mathbb{Z}$-basis $\{1,\zeta_5,\zeta_5^2,\zeta_5^3\}$ of $\mathcal{O}_5$; cf. Lemma 1.99 and the proof of Corollary 1.100.

Step 3: For all $k \in \{1,\dots,t\}$, apply the algorithm presented in the proof of Theorem 3.15 to the image $\Phi[\bigcup_{\ell\in S_k}(\ell-\alpha_k)]$ and obtain the decomposition of the grid

$$G'_k := \bigcap_{\ell\in S_k} \Phi[\ell-\alpha_k]$$

modulo $\mathcal{O}_5$ in terms of the $\mathbb{Q}$-basis $\{1,\zeta_5,\zeta_5^2,\zeta_5^3\}$ of $\mathbb{K}_5$, say

$$G'_k := \dot{\bigcup}_{i=1}^{r_k} C'_{k,i}\,,$$

where $r_k \in \mathbb{N}$. Since all grids G'_k satisfy $\operatorname{card}(G'_k) \le s^2$, Theorem 3.15 shows that this is of complexity $O(s^2)$.

Step 4: For all $k \in \{1,\dots,t\}$, compute the $\mathbb{Q}$-coordinates of the elements of the set $G_k := \Phi^{-1}[G'_k] + \alpha_k$ in terms of the $\mathbb{Q}$-basis $B_{\operatorname{Im}[\mathbb{I}]}$ of $(\Bbbk_5)^3$ and note that

$$G_{\{p_{u_j}|j\in\{1,\dots,m\}\}} = \dot{\bigcup}_{k=1}^{t} G_k = \dot{\bigcup}_{k=1}^{t} \left(\dot{\cup}_{i=1}^{r_k} C_{k,i}\right) \subset G^{\operatorname{Im}[\mathbb{I}]}_{\{u_1,u_2\}} \subset (\Bbbk_5)^3\,,$$

where $C_{k,i} := \Phi^{-1}[C'_{k,i}] + \alpha_k$, $k \in \{1,\dots,t\}$, $i \in \{1,\dots,r_k\}$; cf. Proposition 1.131. One can easily see that, if α,α' are arbitrary elements of $C_{k,i}$ and $C_{k',i'}$, then α and α' are equivalent modulo $\operatorname{Im}[\mathbb{I}]$ if and only if $C_{k,i}$ and $C_{k',i'}$ are subsets of an equivalence class of $G_{\{p_{u_j}|j\in\{1,\dots,m\}\}}$ modulo $\operatorname{Im}[\mathbb{I}]$. This can be done efficiently.

Step 5: Choose arbitrary elements $\alpha \in C_{k,i}$ and $\alpha' \in C_{k',i'}$. Then, $C_{k,i}$ and $C_{k',i'}$ are subsets of the same equivalence class of $G_{\{p_{u_j} | j \in \{1,\ldots,m\}\}}$ modulo $\mathrm{Im}[\mathbb{I}]$ if and only if all $\mathbb{Q}$-coordinates of the difference $\alpha - \alpha'$ in terms of the $\mathbb{Q}$-basis $B_{\mathrm{Im}[\mathbb{I}]}$ of $(\mathbb{k}_5)^3$ are elements of $\mathbb{Z}$; see Corollary 1.100. Thus, one only has to choose one element $\alpha_{k,i}$ of all sets $C_{k,i}$ and use the above fact in order to compute the equivalence classes of $G_{\{p_{u_j} | j \in \{1,\ldots,m\}\}}$ modulo $\mathrm{Im}[\mathbb{I}]$.

By Remark 3.18, there is a constant $c \in \mathbb{N}$ such that, for all instances of the decomposition problem, Step 5 of this algorithm computes the equivalence classes of the grid modulo $\mathrm{Im}[\mathbb{I}]$ in at most c iterations. The inequality

$$\operatorname{card}\left(G_{\{p_{u_j} | j \in \{1,\ldots,m\}\}}\right) \leq s^2$$

(cf. Remark 3.18 again) now completes the proof. □

REMARK 3.20. As in the proof of Theorem 3.15, the proof of Theorem 3.19 indicates that we actually do not need the full strength of the real RAM-model of computation. Again, a Turing machine model that is augmented for algebraic computations suffices.

3.3. The consistency, reconstruction, and uniqueness problems

3.3.1. Planar sets.

DEFINITION 3.21 (PlanarAnchoredConsistency, PlanarAnchoredReconstruction, and PlanarAnchoredUniqueness Problem). Let $u_1, \ldots, u_m \in \mathbb{S}^1$ be $m \geq 2$ pairwise non-parallel directions. The corresponding PlanarAnchoredConsistency, PlanarAnchoredReconstruction and PlanarAnchoredUniqueness problems are defined as follows.

PLANARANCHOREDCONSISTENCY.
Given $s \in \mathbb{N}$ and $p_{u_j} : \mathcal{L}^2_{u_j} \longrightarrow \mathbb{N}_0$, $j \in \{1,\ldots,m\}$, with finite supports whose cardinalities are bounded by s, and a finite set $S \subset \mathbb{R}^2$ with at most s^2 points. Decide whether there is a set F contained in S which satisfies $X_{u_j}F = p_{u_j}$, $j \in \{1,\ldots,m\}$.

PLANARANCHOREDRECONSTRUCTION.
Given $s \in \mathbb{N}$ and $p_{u_j} : \mathcal{L}^2_{u_j} \longrightarrow \mathbb{N}_0$, $j \in \{1,\ldots,m\}$, with finite supports whose cardinalities are bounded by s, and a finite set $S \subset \mathbb{R}^2$ with at most s^2 points. Decide whether there is a set F contained in S which satisfies $X_{u_j}F = p_{u_j}$, $j \in \{1,\ldots,m\}$, and, if so, construct one such F.

PLANARANCHOREDUNIQUENESS.
Given $s \in \mathbb{N}$, a finite set $S \subset \mathbb{R}^2$ with at most s^2 points and a subset F of S. Decide whether there is a different set F' that is also contained in S and satisfies $X_{u_j}F = X_{u_j}F'$, $j \in \{1,\ldots,m\}$.

Further, let $\mathcal{C}$, $\mathcal{R}$ and $\mathcal{U}$ be algorithms for solving PLANARANCHOREDCONSISTENCY, PLANARANCHOREDRECONSTRUCTION and PLANARANCHOREDUNIQUENESS, respectively.

REMARK 3.22. For $m = 2$, the above problems clearly can be interpreted as problems for 0-1-matrices with prescribed zeros and given row and column sums; cf. [**57, 42**]. See also [**41**, Chapter 1] and references therein for a summary. Further, these problems are intimately related with problems concerning bipartite graphs with given degree sequences, the latter being problems which can be dealt with in the framework of network flow theory; see [**31**].

THEOREM 3.23. *For $m = 2$, there are polynomial time algorithms for solving the problems* PLANARANCHOREDCONSISTENCY, PLANARANCHOREDRECONSTRUCTION *and* PLANARANCHOREDUNIQUENESS.

PROOF. For polynomial-time algorithms $\mathcal{C}$ and $\mathcal{R}$ in case of $m = 2$, see [**64**]. There it is shown how to set up a capacitated network that permits a certain flow if and only if the consistency question has an affirmative answer. Moreover, it is shown there that every such flow can be can be interpreted as a solution of the reconstruction problem. Points in the grid correspond to arcs in this network. If one wants to forbid certain positions, one only have to cancel the corresponding arcs. A polynomial-time algorithm $\mathcal{U}$ in case of $m = 2$ can be derived from $\mathcal{C}$ above; compare also [**3, 42**]. □

3.3.2. Cyclotomic model sets. For cyclotomic model sets, the consistency, reconstruction and uniqueness problems from Definition 1.116 look as follows; see also Lemma 1.84.

DEFINITION 3.24 (Consistency, Reconstruction, and Uniqueness Problem). Let $n \in \mathbb{N} \setminus \{1, 2\}$, let $W \subset (\mathbb{R}^2)^{\phi(n)/2-1}$ be a window, and let a star map $.^{\star_n}$ be given (as described in Definition 1.73). Further, let $u_1, \dots, u_m \in \mathbb{S}^1$ be $m \geq 2$ pairwise non-parallel $\mathcal{O}_n$-directions. The corresponding consistency, reconstruction and uniqueness problems are defined as follows.

CONSISTENCY.
Given functions $p_{u_j} : \mathcal{L}^2_{u_j} \longrightarrow \mathbb{N}_0$, $j \in \{1, \dots, m\}$, with finite supports that satisfy $\mathrm{supp}(p_{u_j}) \subset \mathcal{L}^{\mathcal{O}_n}_{u_j}$, decide whether there is a finite set F which is contained in a $\mathcal{M}^{\star_n}_g(W)$-set and satisfies $X_{u_j}F = p_{u_j}$, $j \in \{1, \dots, m\}$.

RECONSTRUCTION.
Given functions $p_{u_j} : \mathcal{L}^2_{u_j} \longrightarrow \mathbb{N}_0$, $j \in \{1, \dots, m\}$, with finite supports that satisfy $\mathrm{supp}(p_{u_j}) \subset \mathcal{L}^{\mathcal{O}_n}_{u_j}$, decide whether there exists a finite set F in a $\mathcal{M}^{\star_n}_g(W)$-set that satisfies $X_{u_j}F = p_{u_j}$, $j \in \{1, \dots, m\}$, and, if so, construct one such F.

UNIQUENESS.
Given a finite subset F of a $\mathcal{M}^{\star_n}_g(W)$-set, decide whether there is a different finite set F' that is also a subset of a $\mathcal{M}^{\star_n}_g(W)$-set and satisfies $X_{u_j}F = X_{u_j}F'$, $j \in \{1, \dots, m\}$.

REMARK 3.25. Note that the parameter n, the window W, the $\mathcal{O}_n$-directions u_j, and the star map $.^{\star_n}$ are assumed to be fixed, i.e., are *not* part of the input. For results on the computational complexity of these problems in the 'anchored' lattice case (and the Turing machine as the model of computation), we refer to [**35, 39**].

3.3.2.1. *Tractability results for cyclotomic model sets with polytopal windows.* As a consequence of Theorems 3.15 and 3.10 we shall see that the standard tomographic algorithms that have been developed for the lattice case can also be extended, in the case of polytopal windows, to the discrete tomography of cyclotomic model sets. More precisely, we shall now show that, for polytopal windows, the problems CONSISTENCY, RECONSTRUCTION and UNIQUENESS for cyclotomic model sets can be reduced to PLANARANCHOREDCONSISTENCY, PLANARANCHOREDRECONSTRUCTION and PLANARANCHOREDUNIQUENESS, respectively.

THEOREM 3.26. *When restricted to polytopal windows, the problems* CONSISTENCY, RECONSTRUCTION *and* UNIQUENESS *can be solved with polynomially many operations and polynomially many calls to* $\mathcal{C}$, $\mathcal{R}$ *and* $\mathcal{U}$, *respectively.*

PROOF. The algorithm for solving CONSISTENCY performs the following steps.

Step 1: Check first the necessary condition that the cardinalities

$$\sum_{\ell \in \mathrm{supp}(p_{u_j})} p_{u_j}(\ell)$$

coincide for each j. If this is the case, say all sums are equal to some $N \in \mathbb{N}_0$, proceed with Step 2. Otherwise the instance is inconsistent.

Step 2: Compute the elements of the equivalence classes G_i of the grid $G_{\{p_{u_1},\dots,p_{u_m}\}}$ modulo $\mathcal{O}_n$, say

$$G_{\{p_{u_1},\dots,p_{u_m}\}} = \dot{\bigcup}_{i=1}^{c} G_i \subset \mathbb{K}_n$$

in terms of their $\mathbb{Q}$-coordinates with respect to the $\mathbb{Q}$-basis $\{1, \zeta_n, \zeta_n^2, \dots, \zeta_n^{\phi(n)-1}\}$ of $\mathbb{K}_n$ (cf. Proposition 1.11). By Theorem 3.15, this can be done efficiently.

Step 3: For all $i \in \{1, \dots, c\}$, compute the $.^{\star_n}$-image $[G_i]^{\star_n}$ of G_i. Note that we consider the star map here as a map

$$.^{\star_n} : \mathbb{K}_n \longrightarrow (\mathbb{R}^2)^{\frac{\phi(n)}{2}-1} .$$

This can be done efficiently. Due to the definition of $\mathcal{M}_g^{\star_n}(W)$-sets, a solution $F \subset G_i$ for our instance must satisfy the condition

$$\exists\, \tau \in (\mathbb{R}^2)^{\frac{\phi(n)}{2}-1} \; : \; [F]^{\star_n} \subset \tau + \mathrm{int}(W) . \tag{3.3}$$

Recall that for $n \in \{3,4,6\}$, condition (3.3) is always satisfied and one can proceed with Step 4. Otherwise, compute, for all $i \in \{1, \dots, c\}$, the set $\mathrm{Sep}_{\mathrm{int}(W)}([G_i]^{\star_n})$. By Theorem 3.10, this can be done efficiently. Note that, for every $i \in \{1, \dots, c\}$, a subset $F \subset G_i$ that satisfies condition (3.3) has the property that $[F]^{\star_n} \subset P$ for a suitable $P \in \mathrm{Sep}_{\mathrm{int}(W)}([G_i]^{\star_n})$. Finally, compute, for all $i \in \{1, \dots, c\}$ and for all $P \in \mathrm{Sep}_{\mathrm{int}(W)}([G_i]^{\star_n})$, the pre-images $S := [P]^{-\star_n}$ of P under the star map. This can be done efficiently. Note that, with the above restriction $n \notin \{3,4,6\}$, the star map is injective.

Step 4: If $n \in \{3,4,6\}$, consider the equivalence classes $S := G_i$, $i \in \{1, \dots, c\}$, having the property that $\mathrm{card}(G_i) \geq N$. Otherwise, consider, for all $i \in \{1, \dots, c\}$ and for all $P \in \mathrm{Sep}_{\mathrm{int}(W)}([G_i]^{\star_n})$, the subsets $S := [P]^{-\star_n}$ of G_i having the property that $\mathrm{card}([S]^{-\star_n}) \geq N$. Then apply $\mathcal{C}$ on each such S. The instance is consistent if and only if $\mathcal{C}$ reports consistency for one of the sets S.

The proofs for RECONSTRUCTION and UNIQUENESS are analogous. □

REMARK 3.27. Note that, in case of CONSISTENCY, the seemingly more natural approach to find subsets $F \subset G_i$ *first* that conform to the X-rays, and check *afterwards* whether (3.3) is satisfied may lead to an exponential running time; see [**8**, Remark 21].

Applying Theorem 3.26 in conjunction with Theorem 3.23 gives the following tractability result:

THEOREM 3.28. *When restricted to two $\mathcal{O}_n$-directions and polytopal windows, the problems* CONSISTENCY, RECONSTRUCTION *and* UNIQUENESS *as defined in Definition 3.24 can be solved in polynomial time.* □

REMARK 3.29. Note, however, that even in the 'anchored' lattice case $\mathbb{Z}^2$ the corresponding problems CONSISTENCY, RECONSTRUCTION and UNIQUENESS are $\mathbb{NP}$-hard for three or more lattice directions (i.e., $\mathbb{Z}^2$-directions); see [**35**].

3.3.3. Icosahedral model sets. For icosahedral model sets, the consistency, reconstruction and uniqueness problems from Definition 1.116 look as follows; see also Lemma 1.109.

DEFINITION 3.30 (Consistency, Reconstruction, and Uniqueness Problem). Let $W \subset \mathbb{R}^3$ be a window (cf. Definition 1.93). Further, let $u_1, \dots, u_m \in \mathbb{S}^2$ be $m \geq 2$ pairwise non-parallel Im[$\mathbb{I}$]-directions. The corresponding consistency, reconstruction and uniqueness problems are defined as follows.

> CONSISTENCY.
> Given functions $p_{u_j} : \mathcal{L}^3_{u_j} \longrightarrow \mathbb{N}_0$, $j \in \{1, \dots, m\}$, with finite supports that satisfy $\mathrm{supp}(p_{u_j}) \subset \mathcal{L}^{\mathrm{Im}[\mathbb{I}]}_{u_j}$, decide whether there is a finite set F which is contained in an $\mathcal{I}_g(W)$-set and satisfies $X_{u_j}F = p_{u_j}$, $j \in \{1, \dots, m\}$.
>
> RECONSTRUCTION.
> Given functions $p_{u_j} : \mathcal{L}^3_{u_j} \longrightarrow \mathbb{N}_0$, $j \in \{1, \dots, m\}$, with finite supports that satisfy $\mathrm{supp}(p_{u_j}) \subset \mathcal{L}^{\mathrm{Im}[\mathbb{I}]}_{u_j}$, decide whether there exists a finite subset F of an $\mathcal{I}_g(W)$-set that satisfies $X_{u_j}F = p_{u_j}$, $j \in \{1, \dots, m\}$, and, if so, construct one such F.
>
> UNIQUENESS.
> Given a finite subset F of an $\mathcal{I}_g(W)$-set, decide whether there is a different finite set F' that is also a subset of an $\mathcal{I}_g(W)$-set and satisfies $X_{u_j}F = X_{u_j}F'$, $j \in \{1, \dots, m\}$.

REMARK 3.31. Note again that the window W and the Im[$\mathbb{I}$]-directions u_j are assumed to be fixed, i.e., are *not* part of the input.

3.3.3.1. *Tractability results for icosahedral model sets with polytopal windows.* In the following, we show that, for polytopal windows and Im[$\mathbb{I}$]$^{(\tau,0,1)}$-directions, also the problems CONSISTENCY, RECONSTRUCTION and UNIQUENESS for icosahedral model sets can be reduced to the problems PLANARANCHOREDCONSISTENCY, PLANARANCHOREDRECONSTRUCTION and PLANARANCHOREDUNIQUENESS, respectively.

THEOREM 3.32. *When restricted to* Im[$\mathbb{I}$]$^{(\tau,0,1)}$*-directions and polytopal windows, the problems* CONSISTENCY, RECONSTRUCTION *and* UNIQUENESS *as defined in Definition 3.30 can be solved with polynomially many operations and polynomially many calls to $\mathcal{C}$, $\mathcal{R}$ and $\mathcal{U}$, respectively.*

PROOF. The algorithm for CONSISTENCY performs the following steps.
Step 1: Check first the necessary condition that the sums

$$\sum_{\ell \in \mathrm{supp}(p_{u_j})} p_{u_j}(\ell)$$

coincide for each $j \in \{1, \dots, m\}$. If this is the case, say all sums are equal to some $N \in \mathbb{N}_0$, proceed with Step 2. Otherwise the instance is inconsistent.

Step 2: Compute the elements of the equivalence classes G_i of the grid $G_{\{p_{u_1},\dots,p_{u_m}\}}$ modulo $\mathrm{Im}[\mathbb{I}]$, say

$$G_{\{p_{u_1},\dots,p_{u_m}\}} = \dot{\bigcup}_{i=1}^{c} G_i \subset (\Bbbk_5)^3$$

in terms of their $\mathbb{Q}$-coordinates with respect to the $\mathbb{Q}$-basis $B_{\mathrm{Im}[\mathbb{I}]}$ of $(\Bbbk_5)^3$ (cf. Corollary 1.100 and Proposition 1.131). By Theorem 3.19, this can be done efficiently.

Step 3: For all $i \in \{1, \dots, c\}$, compute the $.^\star$-image $[G_i]^\star$ of G_i. Note that we consider the map $.^\star$ here as a map

$$.^\star \colon (\Bbbk_5)^3 \longrightarrow \mathbb{R}^3 ,$$

where, for $\alpha \in (\Bbbk_5)^3$, $\alpha^\star$ is again defined by applying the conjugation $.'$ to each coordinate of α. This can be done efficiently. Due to the definition of $\mathcal{I}_g(W)$-sets, a solution $F \subset G_i$ for our instance must satisfy the condition

$$\exists\, \tau \in \mathbb{R}^3 \;:\; [F]^\star \subset \tau + \mathrm{int}(W) . \tag{3.4}$$

Next, compute the set $\mathrm{Sep}_{\mathrm{int}(W)}([G_i]^\star)$. By Theorem 3.10, this can be done efficiently. Note that, for every $i \in \{1, \dots, c\}$, a subset $F \subset G_i$ that satisfies condition (3.4) has the property that $[F]^\star \subset P$ for a suitable $P \in \mathrm{Sep}_{\mathrm{int}(W)}([G_i]^\star)$. Finally, compute, for all $i \in \{1, \dots, c\}$ and for all $P \in \mathrm{Sep}_{\mathrm{int}(W)}([G_i]^\star)$, the pre-images $[P]^{-\star}$ of P under the star map. This can be done efficiently.

Step 4: Consider, for all $i \in \{1, \dots, c\}$ and for all $P \in \mathrm{Sep}_{\mathrm{int}(W)}([G_i]^\star)$, the subsets $S := [P]^{-\star}$ of G_i having the property that $\mathrm{card}([S]^{-\star}) \geq N$. Compute the partition of S with respect to the equivalence relation $\equiv'$ which is defined by

$$x \equiv' y \;:\Longleftrightarrow\; x - y \subset H^{(\tau,0,1)} ,$$

i.e., one has $x \equiv' y$ if and only if x and y lie in a common translate of $H^{(\tau,0,1)}$ in $\mathbb{R}^3$. This can be done efficiently. Then, apply $\mathcal{C}$ to each member of this partition of S. (More precisely, identify all members of this partition of S with subsets of the plane and apply $\mathcal{C}$ to these planar point sets, where also the directions involved and the corresponding functions p_{u_j} have to be reinterpreted accordingly.) The instance is consistent if and only if one of the sets S has the property that $\mathcal{C}$ reports consistency for each member of the above partition of S.

The proofs for RECONSTRUCTION and UNIQUENESS are analogous. □

Applying Theorem 3.32 in conjunction with Theorem 3.23 gives the following tractability result:

THEOREM 3.33. *When restricted to two* $\mathrm{Im}[\mathbb{I}]^{(\tau,0,1)}$*-directions and polytopal windows, the problems* CONSISTENCY, RECONSTRUCTION *and* UNIQUENESS *as defined in Definition 3.30 can be solved in polynomial time.* □

Outlook

Although the results of this thesis give consistent answers to the basic problems of discrete tomography of model sets, there is still a lot to be done in order to finally create a tool that is as satisfactory for the application in materials science as is computerized tomography in its medical or other applications.

First of all, it is an interesting problem to answer the open question which was mentioned in Remark 2.48.

Secondly, since icosahedral model sets based on the face centred icosahedral module $\mathcal{M}_{\mathrm{F}}$ of quasicrystallography (as defined in Section 1.2.3.3) often occur in practice, it is desirable to extend the above results also to this situation.

Thirdly, one might attempt to characterize, for a Delone set Λ in Euclidean 3-space (e.g., a lattice or an icosahedral model set), the sets of Λ-directions *in general position* having the property that the set of convex subsets of Λ is determined by the X-rays in these directions; compare also [**32**, Problems 2.1 and 2.3].

Finally, it would be interesting to have experimental tests in order to see how well the above algorithmic and uniqueness results work in practice. Since there is always some noise involved when physical measurements are taken, the latter also requires the ability to work with imprecise data which, in the case of the consistency or reconstruction problem, can be one reason for inconsistency. In other words, it is necessary to study stability and instability issues in the discrete tomography of model sets in the future.

Bibliography

[1] Agarwal, P.K.; Sharir, M.: *Arrangements and their applications*, in: Sack, J.-R. (ed.) et al.: *Handbook of Computational Geometry*. Amsterdam, North-Holland (2000), 49-119.

[2] Ammann, R.; Grünbaum, B.; Shephard, G. C.: *Aperiodic tiles*, Discrete Comput. Geom. **8** (1992), 1–25.

[3] Anstee, R. P.: *The network flows approach for matrices with given row and column sums*, Discrete Math. **44** (1983), no. 2, 125–138.

[4] Baake, M: *Solution of the coincidence problem in dimensions $d \leq 4$*, in: Moody, R. V. (ed.): *The Mathematics of Long-Range Aperiodic Order*, NATO-ASI Series C **489**, Kluwer, Dordrecht (1997), 9–44. Revised version `math.MG/0605222`

[5] Baake, M.: *A guide to mathematical quasicrystals*, in: Suck, J.-B.; Schreiber, M.; Häussler, P. (eds.): *Quasicrystals. An Introduction to Structure, Physical Properties, and Applications*, Springer, Berlin (2002), 17–48. `math-ph/9901014`

[6] Baake, M.; Grimm, U.: *Bravais colourings of planar modules with N-fold symmetry*, Z. Kristallogr. **219** (2004), 72–80. `math.CO/0301021`

[7] Baake, M.; Grimm, U.; Moody, R. V.: *Die verborgene Ordnung der Quasikristalle*, Spektrum der Wissenschaft (February 2002), 64–74. (in English: *What is aperiodic order?*, Preprint (2002). `http://www.arxiv.org/math.HO/0203252`)

[8] Baake, M.; Gritzmann, P. ; Huck, C.; Langfeld, B.; Lord, K.: *Discrete tomography of planar model sets*, Acta Cryst. A**62** (2006), 419–433. `math.MG/0609393`.

[9] Baake, M.; Huck, C.: *Discrete tomography of Penrose model sets*, Philos. Mag., in press. `math-ph/0610056`

[10] Baake, M.; Joseph, D.: *Ideal and defective vertex configurations in the planar octagonal quasilattice*, Phys. Rev. B **42** (1990), 8091– 8102.

[11] Baake, M.; Kramer, P.; Schlottmann, M.; Zeidler, D.: *The triangle pattern – a new quasiperiodic tiling with fivefold symmetry*, Mod. Phys. Lett. **B 4** (1990), 249–258.

[12] Baake, M.; Kramer, P.; Schlottmann, M.; Zeidler, D.: *Planar patterns with fivefold symmetry as sections of periodic structures in 4-space*, Int. J. Mod. Phys. **B 4** (1990), 2217–2268.

[13] Baake, M.; Lenz, D.; Moody, R.V.: *Characterization of model sets by dynamical systems*, Ergodic Th. Dynam. Syst. **27** (2007), 341–382. `math.DS/0511648`

[14] Baake, M.; Moody, R. V. (eds.): *Directions in Mathematical Quasicrystals*, CRM Monograph Series, vol. **13**, AMS, Providence, RI (2000).

[15] Baake, M.; Pleasants, P. A. B.; Rehmann, U.: *Coincidence site modules in 3-space*, Discrete Comput. Geom., in press. `math.MG/0609793`

[16] Basu, S.; Pollack, R.; Roy, M.-F.: *On the combinatorial and algebraic complexity of quantifier elimination*, J. ACM **43**, No.6 (1996), 1002-1045.

[17] Basu, S.; Pollack, R.; Roy, M.-F.: *On computing a set of points meeting every cell defined by a family of polynomials on a variety*, J. Complexity **13** (1997), no.1, 28–37.

[18] Borevich, Z. I.; Shafarevich, I. R.: *Number Theory*, Academic Press, New York (1966).

[19] Borsuk, K.: *Multidimensional Analytic Geometry*, Polish Scientific Publishers, Warsaw (1969).

[20] Buchberger, B.; Collins, G. E.; Loos, R. (eds.): *Computing Supplementum*. Springer, Wien (1982).

[21] Chen, L.; Moody, R. V.; Patera, J.: *Non-crystallographic root systems*, in: Patera, J. (ed.): *Quasicrystals and Discrete Geometry*, Fields Institute Monographs, vol. 10, AMS, Providence, RI (1998), 135–178.

[22] Chrestenson, H. E.: *Solution to problem 5014*, Amer. Math. Monthly **70** (1963), 447–448.

[23] Conway, J. H.; Sloane, N. J. A.: *Sphere Packings, Lattices and Groups*, 3rd ed., Springer, New York (1999).

[24] Cowley, J. M.: *Diffraction Physics*, North-Holland, Amsterdam (1995).

[25] Darboux, M. G.: *Sur un problème de géométrie élémentaire*, Bull. Sci. Math. **2** (1878), 298–304.

[26] Edelsbrunner, H.; O'Rourke, J.; Seidel, R: *Constructing arrangements of lines and hyperplanes with applications*, SIAM J. Comput. **15** (1986), 341–363.

[27] Edelsbrunner, H.: *Algorithms in Combinatorial Geometry*, EATCS Monographs on Theoretical Computer Science, Vol. 10., Springer-Verlag, Berlin (1987).

[28] Edelsbrunner, H.; Skiena, S. S.: *Probing convex polygons with X-rays*, SIAM. J. Comp. **17** (1988), 870–882.

[29] Esmonde, J.; Murty, M. R.: *Problems in Algebraic Number Theory*, Springer, New York (1999).

[30] Fewster, P. F.: *X-ray Scattering from Semiconductors*, 2nd ed., Imperial College Press, London (2003).

[31] Ford, L. R. Jr.; Fulkerson, D. R.: *Flows in networks*, Princeton University Press, Princeton, N.J. (1962).

[32] Gardner, R. J.: *Geometric Tomography*, 2nd ed., Cambridge University Press, New York (2006).

[33] Gardner, R. J.; Gritzmann, P.: *Discrete tomography: determination of finite sets by X-rays*, Trans. Amer. Math. Soc. **349** (1997), 2271–2295.

[34] Gardner, R. J.; Gritzmann, P.: *Uniqueness and complexity in discrete tomography*, in: [**41**], 85–114.

[35] Gardner, R. J., Gritzmann, P., Prangenberg, D.: *On the computational complexity of reconstructing lattice sets from their X-rays*, Discrete Math., **202** (1999), 45–71.

[36] Gardner, R. J.; McMullen, P.: *On Hammer's X-ray problem*, J. London Math. Soc. (2) **21** (1980), 171–175.

[37] Gähler, F.: *Matching rules for quasicrystals: the composition-decomposition method*, J. Non-Cryst. Solids **153-154** (1993), 160–164.

[38] Gouvêa, F. Q.: *p-adic Numbers*, Springer, New York (1993).

[39] Gritzmann, P.: *On the reconstruction of finite lattice sets from their X-rays*, Lecture Notes on Computer Science, (eds.: E. Ahronovitz and C. Fiorio), Springer, London (1997), 19–32.

[40] Halperin, D.: *Arrangements*, in: Goodman, J.; O'Rourke, J. (eds.): *Handbook of discrete and computational geometry*, 2nd ed., Discrete Mathematics and its Applications. Boca Raton, FL: Chapman & Hall/CRC (2004), 529–562.

[41] Herman, G. T.; Kuba, A. (eds.): *Discrete Tomography: Foundations, Algorithms, and Applications*, Birkhäuser, Boston (1999).

[42] Kuba, A.: *Reconstruction of unique binary matrices with prescribed elements*, Acta Cybernet. **12** (1995), no. 1, 57–70.

[43] Kisielowski, C.; Schwander, P.; Baumann, F. H.; Seibt, M.; Kim, Y.; Ourmazd, A.: *An approach to quantitative high-resolution transmission electron microscopy of crystalline materials*, Ultramicroscopy **58** (1995), 131–155.

[44] Koblitz, N.: *p-adic Numbers, p-adic Analysis, and Zeta-Functions*, 2nd ed., Springer, New York (1984).

[45] Lagarias, J. C.: *Geometric models for quasicrystals I. Delone sets of finite type*, Discrete Comput. Geom. **21** (1999), no. 2, 161–191.

[46] Lang, S.: *Algebra*, 3rd ed., Addison-Wesley, Reading, MA (1993).

[47] Lorentz, G. G.: *A problem of plane measure*, Amer. J. Math. **71** (1949), 417–426.

[48] Moody, R. V.: *Model sets: a survey*, in: Axel, F.; Dénoyer, F.; Gazeau, J.-P. (eds.): *From Quasicrystals to More Complex Systems*, EDP Sciences, Les Ulis, and Springer, Berlin (2000), 145–166. `math.MG/0002020`

[49] Moody, R. V.: *Uniform distribution in model sets*, Canad. Math. Bull. **45** (2002), 123–130.

[50] Natterer, F.: *The Mathematics of Computerized Tomography*, Wiley and Teubner, Stuttgart (1986).

[51] Pleasants, P. A. B.: *Designer quasicrystals: cut-and-project sets with pre-assigned properties*, in: [**14**], 95–141.

[52] Pleasants, P. A. B.: *Lines and planes in* 2- *and* 3-*dimensional quasicrystals*, in: Kramer, P.; Papadopolos, Z. (eds.): *Coverings of Discrete Quasiperiodic Sets*, Springer Tracts in Modern Physics, vol. **180**, Springer, Berlin (2003), 185–225.

[53] Pleasants, P. A. B.; Baake, M.; Roth, J.: *Planar coincidences for N-fold symmetry*, J. Math. Phys. **37** (1996), no. 2, 1029–1058.

[54] Preparata, F. P.; Shamos, M. I.: *Computational Geometry: An Introduction*, Springer, New York (1985).

[55] Radon, J.: *Über die Bestimmung von Funktionen durch ihre Integralwerte längs gewisser Mannigfaltigkeiten*, Ber. Verh. Sächs. Akad. Wiss. Leipzig Math.-Phys. Kl. **69** (1917), 262–277.

[56] Rényi, A.: *On projections of probability distributions*, Acta Math. Acad. Sci. Hung. **3** (1952), 131–142.

[57] Ryser, H. J.: *Matrices of zeros and ones*, Bull. Amer. Math. Soc. **66** (1960), 442–464.

[58] Salem, R.: *Algebraic Numbers and Fourier Analysis*, D. C. Heath and Company, Boston (1963).

[59] Schlottmann, M.: *Cut-and-project sets in locally compact Abelian groups*, in: Patera, J. (ed.): *Quasicrystals and Discrete Geometry*, Fields Institute Monographs, vol. **10**, AMS, Providence, RI (1998), 247–264.

[60] Schlottmann, M.: *Generalized model sets and dynamical systems*, in: [**14**], 143–159.

[61] Schwander, P.; Kisielowski, C.; Seibt, M.; Baumann, F. H.; Kim, Y.; Ourmazd, A.: *Mapping projected potential, interfacial roughness, and composition in general crystalline solids by quantitative transmission electron microscopy*, Phys. Rev. Lett. **71** (1993), 4150–4153.

[62] Schwarzenberger, R. L. E.: *N-dimensional crystallography*, Pitman, San Francisco (1980).

[63] Sloane, N. J. A., (ed.): *The Online Encyclopedia of Integer Sequences.* `http://www.research.att.com/~njas/sequences/`

[64] Slump, C. H.; Gerbrands, J. J.: *A network flow approach to reconstruction of the left ventricle from two projections*, Comput. Graphics Image Process. **18** (1982), 18–36.

[65] Steurer, W.: *Twenty years of structure research on quasicrystals. Part I. Pentagonal, octagonal, decagonal and dodecagonal quasicrystals*, Z. Kristallogr. **219** (2004), 391–446.

[66] Washington, L. C.: *Introduction to Cyclotomic Fields*, 2nd ed., Springer, New York (1997).

[67] Weyl, H.: *Über die Gleichverteilung von Zahlen mod. Eins*, Math. Ann. **77** (1916), 313–352.